PRAISE FOR THE TRUTH

"A delightful and insightful guide to know your truth and live in joy."

—John Gray, Ph.D., #1 bestselling relationship author of all time: Men are from Mars, Women are from Venus, and whose books have sold over 50 million copies in 50 different languages around the world.

* * * * *

"I like the spirit of this book, *Truth about Life*. It is refreshing that somebody is talking about truth in this age of scientific materialism, relativity of truth, and Fox news … Read this book; it will inspire you."

—Amit Goswami, Ph.D., world-renowned theoretical Quantum Physicist

* * * * *

"Lilly Koutcho brings wisdom to our ears and life to our years as she shares through her teachings the strong distinction between facts and spiritual truths. She's more than an author—she's a living example of true change. Everyone must read this beautiful story!"

—Rev. Dr. Temple Hayes, internationally recognized motivational speaker, radio host for Unity FM, and author of several books.

* * * * *

"Dr. Koutcho has written a must-read primer for anyone seeking the highest and purest truth and love. Read this book *and learn from one of the best.*"

—*Dr. Rajiv Parti, MD, former founder of the Pain Management Institute of California, former Chief Anesthesiology at Bakersfield's Heart Institute Hospital, and author of the best-selling book Dying to Wake Up.*

* * * * *

"Every thought is a prayer, affecting every aspect of your life. Here, indeed, is THE TRUTH!"

—*Norman Shealy, M.D., Ph.D., founding president of the American Holistic Medical Association*

* * * * *

"*The Truth About Life* takes on the task of explaining a greater truth through the lens of the physical, philosophy, and religion. For those who enjoy ancient insights of truth, this is a must-read!"

—*Jody Long, attorney, licensed in Washington, New Mexico, Louisiana, and the Navajo Nation. Author of several books and webmaster for the Near-Death Experience Research Foundation (NDERF).*

ALSO BY DR. LILLY KOUTCHO

BOOKS:

Healing & Abundance Affirmations: A Gift From Your Angels

A Prayer Book for Realizing Inner Peace

THE TRUTH ABOUT LIFE

Revelations Never Told before
in the History of Humanity

DR. LILLY KOUTCHO

Copyright © 2020 by Lilly Koutcho

www.integrityhealing.org

All rights reserved. No part of this book may be used or reproduced by any means—graphic, electronic, or mechanical, including photocopying, recording, taping, or by any information storage retrieval system—without the written permission of the author, except in the case of brief quotations embodied in critical articles and reviews.

The intent of the author is only to offer general information to help you in your quest for emotional and spiritual welfare. The information given in this book should not be treated as a substitute for professional medical advice; always consult a medical practitioner. Any use of information in this book is at the reader's discretion and risk. Neither the author nor the publisher can be held responsible for any loss, claim, or damage arising out of the use, or misuse, of the suggestions made, the failure to take medical advice, or for any material on third-party websites.

"Three things cannot be long hidden: the sun, the moon, and the truth."

—Buddha

"Many people, especially ignorant people, want to punish you for speaking the truth, for being correct, for being you. Never apologize for being correct, or for being years ahead of your time. If you're right and you know it, speak your mind. Speak your mind. Even if you are a minority of one, the truth is still the truth."

—Mahatma Gandhi

TABLE OF CONTENTS:

Prologue i
How to Read This Book iii
Preface iv
Introduction vii

CHAPTER 1: Who Are We? 1

The Lion Who Became a Sheep 1
Who Are We? 4
Oneness Principles 7
Illustrative Analogies and Metaphors of "Who We Are" 13
Why Are We Here on Earth? 14
Where Do We Come From and What Are We Doing Here? 16

CHAPTER 2: The Physical Body 18

What Can You Tell Us about the Physical Body? 18
The Chakra Systems 27

CHAPTER 3: Health and Disease 32

Why Do We Get Sick, and Where Do Diseases Come From? 32
Examples of Health Issues and their True Meanings 45

The Biggest Unknown Truth about Diseases 49

Children and Health Issues 54

CHAPTER 4: Thoughts, Beliefs, Creation Of Reality 59

Thought, Beliefs, and How to Release Negative Thought
Patterns 59

Ways to Change Negative Thought Patterns 66

Affirmations as a Powerful Tool 76

CHAPTER 5: The Ego 83

What Is the Ego? 83

Various Roles Often Taken by the Ego 86

Why Do We Have Ego? 91

Other Characteristics of the Ego 92

Identification with the Ego 94

Getting out of the Ego 98

CHAPTER 6: Reincarnation, Karmas, Dharma, and Choices Before Birth 105

Reincarnation 105

Do Karmas Exist, and How Can You Explain Them? 106

Dharma or Experiences Elected before Birth 107

CHAPTER 7: Death, Life After Death, Heaven 113

CHAPTER 8: Soul Age, Soul Group, Soulmates 119

CHAPTER 9: Religion 123

Subchapter 1: The Truth about Religion 124

What Is the Truth about the Various Religions? 124

The Blind Men and the elephant 126

Subchapter 2: God 131

Who or What Is God? 131

Clarification of Some Misconceptions about God 135

Subchapter 3: Jesus 139

Is Jesus the Son of God? 139

Is Jesus the Only Son of God? 139

What Is the Meaning of the Death of Jesus on the Cross? 140

Did Jesus Resurrect from Death? 142

Did Jesus Walk on Water? 143

Is Jesus Coming Back? 143

Subchapter 4: Islam and Mohamed 145

What Can You Tell Us about Mohamed and the Birth of the Islamic Religion? 145

Did Islam Originate from the Pure Light and Love of God? 146

How Has Islam Become What It Is Today? Why the killing? It is Completely out of Love. 146

Other Truth Detecting Tips: 149

Subchapter 5: Prayer **151**

What Can You Tell Us about Prayers? 151

Subchapter 6: Bible Truths Revealed **161**

What Can You Tell Us about the 10 Commandments? 161

What Is the Truth about Tithes and Tithing? 163

Who is Mother Mary? 168

What Can You Tell Us about the First Man, Adam, and Creation of the Earth? 169

Did God tell Abraham to Sacrifice His Son? 170

Who Are the Chosen People? 170

What Can You Tell Us Regarding the Names or Attributes of God Such As the Ones Mentioned in the Bible? 171

Can You Explain Some of the Concepts Described in the Book of Revelation? 172

Subchapter 7: The Notion of Hell **174**

What Can You Tell People about the Existence of Hell? 174

CHAPTER 10: Relationships, Sexuality, Marriage, Divorce, Unfaithfulness, Celibacy, Abuse 179

Sexuality, Marriage, Divorce, Remarriage, Celibacy 179

Infidelity, Polygamy, Abuse 190

Self-love 193

CHAPTER 11: Gender, Transgender, Personality Traits, Feelings, and Emotions 198

Gender, Transgender 198

Personality Traits, Feelings, Emotions 201

CHAPTER 12: Children..204

Indigo Children, Crystal Children 208

CHAPTER 13: The Truth About Mass Media, Television, and Celebrity Magazines 211

CHAPTER 14: Tragedy, Disease, Natural Disasters, People with Disabilities 220

CHAPTER 15: MISCARRIAGES, ABORTION, SUICIDE, DRUG ADDICTION...226

CHAPTER 16: Future and Fate of the Earth, Dimensional Shift, and Graduation...231

CHAPTER 17: Clarification of Some Big Misconceptions and Lies About Life...237

CHAPTER 18: Angels, Archangels, Ascended Masters........249

Archangels 252

Ascended Masters..268

The new generation of ascended masters: 280

CHAPTER 19: Other Planets, Galaxies, and Beings from Other Dimensions 311

CHAPTER 20: Healthy Living, Food, Exercising, Sun, Grounding, Aging 316

Food and Physical Activity 316

Tips and Techniques for Healthy Eating Habits to Increase
Your Vibration 323

Sunlight, Electronic Wavelengths, Grounding 326

CHAPTER 21: Money, Manifestation, Abundance 330

The Truth About Money 330

Steps for Manifesting Abundance and Money 334

CHAPTER 22: The Seven Pillars for Success in Life 347

Acknowledgments 361

Postface 362

References 363

About the Author 364

PROLOGUE

The information contained within this book is channeled through the vehicle of Lilly by the Archangel support team. Through Christ Consciousness and through my celestial team, I present to you these truths of Light and Life.

How Was This Book Written?

This book was written with the help of my angelic team comprising Archangel Gabriel, Archangel Uriel, Christ Consciousness (formally known as Jesus), Saint Augustine, and my celestial team. I often sat, said a prayer, asked for guidance, and then I would start writing. Many times I received guidance in my dreams or through my intuition and would write them down. The great majority of the book was written through direct channeling, where I will receive direct information from my guides and angels about the various topics presented in this book. When my friends, relatives, and other people learned that I was writing a book titled "The Truth about Life," they started to ask me many questions. I often redirected these questions to my angelic heavenly team, who would answer them and give helpful explanations. Some of these questions are included in this book. At the end, I compiled all the materials into chapters so that they are easily accessible, understandable, and helpful for everyone.

The channeled material will help you evolve, and you will feel activation as you read them. If it resonates, and you connect to this material, read on. If it does not, that is OK.

The intent of this book is to help you find the truth, to live in peace, and to ease your journey on earth. When you are done with it, move to the next helpful information, to the next activation, because every soul is evolving. It does not matter where come from or what has happened in the past; the only thing that matters now is that you are here reading these pages. You can choose today to implement these principles and practices, these truths.

Behind these words, there is the energy of Truth. Look for yourself, feel for yourself; then you know that you are walking in Light and Truth. As you read these words, you will feel its truth, and that speaks for itself.

HOW TO READ THIS BOOK

As you progress your way through this book, your understanding of truth will increase, and you will eventually integrate the concepts of truth embedded in this book. Every time you read this book, you are planting the seedlings of truth in your garden of consciousness, and you are enriching the soil of your consciousness with fertilizers of truth. Whoever reads these words will have planted the deposits of the seeds of truth in their garden of life. The more you read this book, the more you will ingest the seedlings of truth, and the more your garden will grow.

I suggest that you read this book numerous times. If you read it just once, that is okay, but you may not harvest the full fruits of truth from the tree of your garden of life. However, every time you read it, it will become more meaningful to you, more concepts will become clearer to you, and it will unfold for you. Then the seedlings of truth will ignite. If you ingest it and read it enough, you will become the truth that lies within these pages.

With love and gratitude,

Peace,

Lilly

PREFACE

What is the truth about life? What if everything we think we know about life is not true after all? What if everything you have been taught by society, your parents, the media, your religion regarding life is not 100% accurate? Have you ever questioned your beliefs? What if you are seeing life through a spotted and cloudy lens that distorts and discolors the reality of everything you see? Have you ever thought that you knew something or someone, only to discover later that you didn't know that much about them after all?

Thousands of years ago, when the Greek philosopher Pythagoras claimed that the earth was round and not flat, it was a foreign and ridiculous idea for people of his time and untrue for them because they falsely believed that they detained the truth. Today, we know for a fact that the earth is indeed round.

Likewise, at the beginning of the 19th century, when the Wright Brothers were attempting to fly the first airplanes, they were mocked because people of their time thought that it was not humanly possible or true for humans to rise up in the air in metallic flying vehicles that we know today as airplanes.

The question then is what if many things that we think we know about life are not true after all? What if we are in the same closed-minded belief that life is a certain way (flat), while indeed it is not true, and life is another way (round) after all?

THE TRUTH ABOUT LIFE

Is it possible that what we know about life, death, health, the cause of disease, abortion, tragedies, God, natural disasters, relationships, homosexuality, religion, etc., is not true or is outdated information? What if we are out of touch with the truth once again?

This book provides answers to some of the most important and pressing questions in life and fearlessly examines these subjects through the lens of truth and love. In this book, you will receive clear and concise answers to the hottest questions that have been debated by humanity for millennia, throughout the history of the earth, and have divided humans, as these subjects were labeled as "too sensitive," "taboos," "unknown," "falling into relativity." Now the time has come for the truth to be revealed.

Many think that there are several truths or that the truth is relative and changeable. However, the truth is the truth, with no need for explanation or justifications. For instance, the earth cannot be round and flat at the same time; it is either round or flat. Similarly, the truth about a particular subject cannot be one way sometimes and another at other times. Anything in between is either partial truth, false, or a versatile interpretation of the truth. A partial truth is not the truth because it does not represent the whole truth and, therefore, is nothing but a disguised misinterpretation of the truth, and subsequently a lie.

In this book, you will be presented with the truth, the absolute truth. This is not a book about religion, even though it will reveal truth and clarity on some of the religious practices that have mined the earth for millennia. This is not a book about the denunciation of manipulations that we, human beings, have been victims of on Earth, even though it will reveal some truth and clarity about some of these topics. This is not a book about God; therefore, you do not have to believe in God or anything in order to benefit from the truths provided in this book.

Whether you believe in something or whether you are a self-proclaimed atheist who has rebelled against the word "God," this book provides information that transcends all beliefs and religions.

The promise of this book is to provide you with the truth and the correct information so that you don't walk in the darkness wandering and lost in life. This book reveals the truth about life; the reason why we are here on Earth; the truth about relationships; the truth about how we create our reality; the truth about the voice that speaks in our minds; the truth behind the media system, its real purpose and unseen effects on us; the truth about the real causes of diseases and how to heal; the truth about money and how to manifest wealth and abundance in your life; the truth about tragedies, natural disasters, the future of the earth; about "who or what" we call God and religions; the truth about why some people are homosexual or have different sexual inclinations; the truth about the sun's rays and its real effect on the body; the truth about Jesus, the prophet Mohamed, etc.

INTRODUCTION

My Story

I have learned the truths about life the hard way. I wish someone had taught me these truths or that I was presented with books early on in my life, but, unfortunately, that was not the case. I was born in a familial environment where these concepts were foreign ideas. Therefore, I learned through the process of trials and fails, and I have learned the truth about life mostly through painful experiences, heartbreaks, humiliations, and challenges. Since this book is about revealing the truth about life, it is appropriate that I tell you the truth about myself and how I came to discover these truths about life.

How I Discovered the Truths about Relationships

I born in a small country in West Africa, Togo, and was raised by a single mom with my two other siblings. My father, who was financially well established and rich by the norms of our society, cared little about my two sisters and me and did little to provide for our needs. In fact, it was through constant arguments with my mother that he reluctantly paid for our school tuitions, and yet some years he would decide not to pay anything at all because someone or something had upset him. My father was a polygamist, had more than a dozen children that he did not support either. Even though he had the financial means to help, he simply did not care. I could not understand how someone could behave like that. I grew up with a sense of rejection, bitterness, and stuffing anger toward my father. My mother with her meager secretary salary of approximately $80 a month did not have enough money to feed us, to pay for rent, and to care for us. We did not have a stable place to live. Therefore, we lived in a tiny bedroom in my grandfather's home; he generously allowed us to stay without paying rent. My two sisters, my mother, and I shared the same bedroom, with my mother sleeping on a net on the ground, letting my sisters and me sleep in the only bed that we had. From there, we later moved to live in one of my uncles' home, and after familial issues, we were asked to move out. Not knowing where to go, my mother and sisters moved to live with one of my aunts.

It was painful not having a place that I could call home. Throughout my childhood and teenage years, I knew that we were not like the other children. I was often ashamed of our situations and sometimes tried to hide our poverty and familial situations from my friends. I was embarrassed when my friends asked, "We have never seen your father; who is your father?" I did not know what to tell them. I felt shamed and embarrassed. However, my mother was struggling to raise us to the best of her ability.

I often asked myself, "Why did I come into the world in such unfavorable circumstances, whereas my cousins and friends live with their mothers and fathers who take great care of them?" Later, after graduating from high school, I was able to move to Europe and later into the United States for college studies. My frustration toward my father grew when I became an adult and realized that every time he called me over the phone, it was only to ask me for money.

I was the youngest child and had two other sisters, and my relationship with my mother, though better than my relationship with my father, was not smooth either. My mother is a well-intentioned woman who was very devoted to the success of her children and did all she could to help us succeed in life. She decided to see her children succeed and not see them having the same fate in life as herself. However, like every human being, she had her strengths and weaknesses. In fact, my mother openly had her favorite child, who is my oldest sister. My other sibling and I were never good enough, not the best, always made mistakes, not as good as my oldest sister, and this was not something that my mother was shy about but openly shared to whoever she knew. The best foods, the best spot in the house, the best of everything was reserved for my oldest sister. My oldest sister even received nearly double the monthly allowance we received. This was just the way it was, and it was all normal for my mother. It was painful to be relegated to the place of "underdog child" and even more painful to know that no matter what I did, I would never get the attention, the special treatments, and favors that my oldest sister received from my mother. It was almost as if my other sibling and I were outcasts in our own family. This open preference of my mother led me to feelings of frustrations, rejection, and unfairness that I unconsciously carried into my adulthood.

Later, I went from one abusive relationship to another. I realized later that I was looking for someone who would love and accept me as I am but only faced disappointments, abandonment, and the same feelings of rejection that I faced in my childhood. Little did I know that the shadows of my parents were following me in my adult relationships. It was later that I discovered the truth about relationships and how carrying unresolved pains and heartbreaks from previous relationships and even from childhood can draw to us similar relationships and partnerships like magnets through the universal law of vibration/attraction. It was as if a veil fell off from my eyes, and when I started to honestly examine my love life and relationships, I realized to my great astonishment that indeed the same patterns and situations kept repeating, and my ex-partners did exhibit the very same characteristic traits that I hated in my parents. Again and again, I faced the same deep sorrows and feelings of not being good enough, not being loved for who I was; the same rejection, abandonment and betrayal that I felt in my childhood. I learned later through communication with my guardian angels and ascended masters that my childhood experiences, my own self-esteem issues, the resentments that I was carrying toward my parents were the elements that attracted and shadowed my past love relationships and friendships and were one of the causes of the relationship problems that I faced in life. Healing my childhood heartache, pains, and traumas, forgiving and letting go of old resentments was what would heal my relationships and help me to attract new, healthy relationships. What an eye opener that was for me. I realized later the truth of that wisdom because, as I started my healing and forgiveness process, my relationships started to change, and new, loving people started to come into my life. Even my relationship with my parents started to improve as we reconnected.

How I Discovered the Truth about Health

I was living in the Midwest of the United States, was in pharmacy school, and had an 18-month-old child to care for mostly alone because my husband was working in another state at the time and only came back home on weekends. I also had a part-time job as a pharmacy technician in a hospital at the time. Between the heavy demand of pharmacy school, the challenge of caring for my child, and the stress of working, I was overwhelmed, but I needed to keep going. I had invested nearly five years in the pharmacy curriculum and had several thousands of dollars of student loan. Therefore, I needed to keep going. Giving up was not a choice for me. I had several health issues my whole life, but my health condition was getting worse, and I ended up taking several pills a day to treat the many health problems that I had—such as chronic headaches, chronic spasmodic allergy, dysmenorrhea, heartburn, hemorrhoids, chronic sinus issues, wrist pain, frequent infections, and so on. I was also drinking a ton of coffee every day so that I could have enough energy to get through the days. It got to a point where I was taking nearly 17 pills a day and drinking nearly a liter of coffee daily just to keep up with life.

During that period of my life, I was dealing with other stresses as well, notably relationship issues with friends and family. I was angry, bitter, filled with resentment, and enraged about the many problems and issues that I had in my life, and yet I was determined to succeed and reach my goals at all costs, even if that meant living on pills and coffee daily.

One day, I literally had a mental breakdown after a serious family issue with my mother. I reached the bottom at that point and knew that something needed to change. I did not know what needed to change at first, but I came to realize that what needed to change was not the other people, my life circumstances, or my family but me. What a shock! I realized that I had no power over other people and what other people were doing and that the only person that I had control over was me and my reaction toward others. I could no longer live that way—filled with anger, resentment, bitterness, and all the heaviness of the past that I was carrying. It was then that I decided to change myself and to forgive everyone for everything. However, at the beginning, my decision to change, to forgive others, and to become more loving was just a choice that I had made in order to find peace in my mind and heart. This decision had nothing to do with my health because, at that time, I knew nothing about the truth about health and disease and just wanted to have peace in my own life by forgiving people and releasing the past.

I then went on a long journey, where I was applying forgiveness principles that I had learned from various books. I needed to forgive everyone and myself and to let go of bitterness, anger, and childhood traumas. I also decided to implement other new, positive habits in my life as well such as holding a daily gratitude journal, trying to do something nice for someone every day, even if it was as simple as opening a door for a stranger or giving a dollar bill to a homeless person. I also started to meditate daily, starting with five minutes of meditation daily and slowly increasing the time to 10, 15, and 30 minutes. During that time, I also decided to let go of all negativities in my life, including cutting off the time spent watching negative TV programs and dramas. I also started to stay away from drama-filled conversations and people who are negative in nature.

To my greatest astonishment, I started to notice that, as I was forgiving and changing my life, my health issues were improving. Something was happening that was beyond my understanding. My heartburn issues disappeared one day. Then the terrible recurrent headaches, hemorrhoids, dysmenorrhea, wrist pain, recurrent infections, and all of my health issues were disappearing one after the other. Even the chronic sinus allergy that I had for over 15 years simply disappeared one day. "What is happening?" I often asked myself. I was healing! I was getting off of the pills one by one, and one day, I became completely disease-free and pill-free, and I was healthier than ever before. What I did not know at the time was the truth about diseases and that it is all a matter of vibration and that diseases are related to negative toxic emotions and feelings that we hold. What I failed to realize at the time was that the helpful habits that I was implementing—forgiveness, meditation, practicing loving kindness, becoming a vegan, cutting off all negativities from my mind and my life, including negative TV programs and toxic relationships—were increasing my vibration and were the very things

THE TRUTH ABOUT LIFE

that were helping me to heal. I came to a point where I spent years without even having a slight headache or cough. Yes, it is possible to live completely healthy and disease-free.

I learned later the truths about health and diseases and how the body is malleable, depending on the frequencies of thoughts and emotions bestowed upon it from my celestial team of archangels and ascended masters, and these truths have completely changed my life forever. These truths are presented in this book.

How I Discovered the Truth about Diseases

Since I was young, I often had premonitory dreams about events or people, and these events somehow happened later. Everyone in my family knew about the visions and messages that I often received in my dreams, and they ended up taking them seriously as they had learned with time that these dreams were quite accurate. My family and friends simply knew this weirdness about me, and it was all fine with them. They often asked me, "I heard that you had a dream about such and such. Can you tell us more about it?" Oftentimes, I would receive helpful guidance in my dreams and acted upon them and made decisions based on my dreams, and later, these decisions ended up becoming lifesaving decisions. For instance, one morning, I woke up and called my sister, who was a manager in a big corporate bank in California at the time, to tell her to be very diplomatic and careful in firing one of her employees because I received messages in my dreams that the employee was planning to create serious issues and probably take legal actions, which were going to put my sister in some serious troubles. My sister was shocked and announced to me that she was indeed in the process of firing one of her employees and was planning to have a meeting with that employee that week. However, following the guidance received in my dreams, my sister postponed the meeting, changed her plans, took other extra precautionary steps and involved other supervisors in the firing process, and everything finally went smoothly for everyone involved. It was so for years; I would receive a message, guidance about people, situations, and events in my dreams, and it was just a normal part of my life and who I was.

Things took another turn when I started to feel people or intuitively know what was going on in their lives or even hear their thoughts as they thought them. In fact, during my last year of pharmacy school, which was a clinical year, I noticed that I was receiving information about patients while I was in clinical settings with them. For instance, a patient would come to consult for anxiety and trouble sleeping, and while talking with the patient, I would receive information that she was sexually abused in her childhood, and this was the real cause of her anxiety and health issues. I knew without a doubt that the information I was receiving was accurate without knowing how or where this information was coming from. I received several messages and guidance about patients often linked to the emotional distress that they were going through, their marital issues, the struggle to find their place in life, their desires to die, the feelings of unworthiness and not being enough, while medically they simply came into the clinics with various diagnoses such as diabetes, strokes, anxiety, heart failure, and so on. I knew then that there was something more, a deeper truth about health and disease that we do not know. I started to notice that the apparent diseases that my patients had were just the surface manifestation of something deeper related to their emotionality, their deepest pains, their negative thought patterns, and painful childhood traumas and experiences. By connecting the dots, I discovered the truth about diseases and that truth was quite surprising for me as a scientist, biologist, and pharmacist.

THE TRUTH ABOUT LIFE

How Did I Discover the Truth about God?

Born into a Christian tradition, I was particularly interested in church and the Bible. As a child, I believed that God is a loving God, but I had many questions that I did not understand and which intrigued me. *Why did a loving God ask Abraham to sacrifice his son to Him, for instance?* It seemed like a terrible thing to ask someone to kill his own son with a knife, even if it was just a test. So, why did a loving God ask Abraham to do such a thing? It did not make any sense to me. *Why does God test us anyway when He already knows what is in our hearts?* These are the questions I often asked myself. However, as soon as I questioned these things, I felt deep fears. Who was I to question God anyway? What if these questions brought the wrath of God upon me? *After all, isn't He God, and can't He do whatever pleases Him?* These deep fears prevented me from questioning further. Nevertheless, these questions and incongruence remained in my mind, and I often pondered on them for years.

Why did God say, for instance, "An eye for an eye and a tooth for a tooth," and only to later command, "Do not repay evil with evil or insult with insult, but repay evil with blessing," and "Love your enemies, pray for your enemies"? Does God change His mind over time? However, at the same time, I was told that God never changes. These inconsistencies and discrepancies intrigued me, and my questions and curiosity only grew with time. My whole life, I was literally in search of answers for the truth. My quest for answers to these questions led me to discover the truth about God, the truth about life and who we truly are. These truths have changed the way I see people and my perception of life for good.

After discovering these truths about life, I started to apply their principles to change and better my own life, and the result has been amazing. During my discussions with others, when people realized how I had changed, they often asked me several questions regarding the truth on various topics of life. Many have told me that our conversations and the answers helped them to live better and to be more at peace. Some with whom I shared this information started to apply the principles of these truths in their lives and informed me later that they were healed, happier, and more peaceful. Some suggested that I share these conversations with others and write them down, and that was how this book came to life.

My prayer is that this book truly helps you to get in touch with the truth, live healthier and happier, and helps you to find peace in life.

CHAPTER 1

WHO ARE WE?

"There's an old saying that if a lie is told often enough, it becomes the truth. Actually, it doesn't. What happens is that people simply start believing that it's true."

–Bailey Jackson

The Lion Who Became a Sheep

A retelling of an ancient traditional Indian story

Once upon a time, there was a huge lioness pregnant with a lion cub. The pregnant lioness was hungry and ventured deep into the jungle, away from the other lions, in her quest for prey to eat. After walking long hours without any success, she fell asleep beneath a tree in the jungle. Suddenly, she heard a bleating sound, which woke her up. She opened her eyes and discovered, to her amazement, a big flock of sheep walking by. Starving, she ran after the flock of sheep, which, in turn, started to run away in panic at the sight of the lioness. The lioness successfully captured one of the sheep and took it away into the jungle for her meal. She didn't even realize that in the midst of the excitement and chaos engendered by the pursuit of the sheep, she had given birth to her lion cub.

After the paralyzing fear and chaos was over, the sheep mourned with sorrow when they realized that one of their friends was taken away by the lioness. In their lamentation and cries, they discovered to their surprise that there was a lost, innocent, helpless lion cub among them. One of the sheep felt compassion for the lion cub and decided to adopt him as her own. The orphaned lion lived among the sheep, was raised like a sheep, ate grass instead of meat, and bleated like a sheep instead of roaring like a lion. This vegan lion grew up like a sheep, called his adoptive sheep mother "Mammy," behaved like a sheep, and was well loved by his new community of sheep.

He lived among the sheep until one day a mighty king lion was hungry and ventured far into the jungle in search of food. He saw the flock of sheep and started running after them for prey. The king lion was shocked to see that among the flock of sheep there was a huge lion who was bleating like a sheep and running away from him with terror. Concerned and disturbed at seeing another lion so scared of him and behaving like a sheep, he decided to run after the sheep-lion to see what was going on and what was the matter with him. The king lion was faster than the now grown-up sheep-lion, and he successfully captured him. Paralyzed with fear, the sheep-lion pleaded, "Bleat, bleat, bleat. Have mercy on me, oh lion. Do not kill me. I am just a little sheep who wants to live."

The king lion, shocked by what he heard, shook him and said, "Wake up! What is wrong with you? You are not a sheep; you are a lion like me; you are my brother. Why are you so afraid of me?"

However, the sheep-lion, still frightened, and with his eyes still closed, said once again to the king lion, "Bleat, bleat, bleat. I am just a little sheep. Have mercy on me and do not kill me. Please let me go."

Determined to show the sheep-lion the reality of who he was, the king lion dragged him to a stream of water near the jungle and ordered, "Open your eyes and look at your face reflected in the water and take a look at my face; they are the same. You are not a sheep. You are a lion like me, and lions do not bleat but roar."

After seeing his face in the reflection of the stream of water, the sheep-lion became convinced that he was indeed a lion and not a sheep. Under the recommendation of his new friend, the king lion, the sheep-lion tried to roar but only succeeded in producing a feeble bleat-like roar. However, after several other attempts under the coaching of his new friend the king lion, the sheep-lion finally succeeded in making a real lion-like roar. Now friends, both lions joyfully went across the field, celebrating their new friendship…

This story perfectly illustrates how many of us have lost touch with the reality of who truly we are. Though we are divine powerfully lion beings, we have forgotten the truth of who we are, and we falsely think that we are sheep because we are born into this sheep-like earth life. Raised and conditioned by our society to bleat like sheep and frightened with fear, untruthful thoughts, unworthiness, not being good enough, lack, scarcity, and terrified of the idea of death, we have forgotten that we are, in fact, divine, immortal, eternal beings of love, powerfully and wonderfully made like lions.

This chapter is about remembering the truth of who we really are by looking at the stream of water—mirror of truth—as a way of refreshing our long-lost memories buried beneath our conditioning by society, our parents, educations, cultures, and religious training.

Who Are We?

The fundamental truth that we need to know is who we truly are. This truth is the cornerstone, the foundation from where all else flows. You can build a nice home, even an amazing royal castle, but if the foundation of your home is built with rotten wood, your effort is vain, and it will be just a matter of time before your home or your castle collapses. Therefore, we must first come back to the simple basic truth and know who we truly are. This book is dedicated to the truth about life, but before we can dive in to reveal the truth about life and about various facets of life, we must first know the truth about ourselves, who we truly are, and the very reason why we are here. We have lost touch with the reality of who we are, and this is the cause of all of the confusion, suffering, and pain that many of us experience in this life.

The truth is you cannot know who you really are by thinking about it or using words. You can read books and ponder on it, but none of this will tell you who you are. This truth must be perceived, realized, and experienced. You cannot get there through your brain by thinking about it.

Since thinking cannot help you to know the truth about who you are, the first step naturally will be to cease all thoughts—stop thinking for a while. You can do this by sitting still in silence. When you do so, you will notice that something is aware that you are thinking. Therefore, there must be a part of you, an Intelligence that is back behind the thinker. It is called Consciousness, Aliveness, Presence. This Consciousness is who you truly are. You feel it immediately when you are quiet, silent, and still your mind. By quieting the mind, you will begin to sense that there is something beyond all your life experiences, beyond all the stories that you have about yourself, beyond all the noise in your head, and this "something" is who you truly are, and it is "what" is aware that you are here. This "something" is a Presence, an Inner Light, a peaceful Consciousness. That "something" which cannot be named is who you truly are.

Though I would like to warn you that if you try to sit in silence by quieting your mind to feel that Inner Presence of who you truly are, you will notice how quickly the "mind" will interject and give you thousands of reasons why you should not sit quietly, reasons why you have something better to do than sit quietly. This is the "ego". The ego is the part of the mind that thinks all the time—which we falsely believe to be us. This ego, the thinking mind, is one of the biggest challenges that we face as human beings. Later in this book, I will expand on what the ego truly is. For now, the truth is that there is something within you that is observing all of this. If you are able to notice and say, "I am thinking," the questions to ask yourself are: Who is the 'I' that is aware that it is thinking and who is 'the thinker'? Just as the eye cannot see itself, if you are able to notice that you are thinking, there must be a true you who is noticing that you are thinking and who is different from the thinker (the egoic mind). Who is the "observer"? Who is the "Intelligence" that is able to look, to perceive that you are thinking, and to notice all of these? This "Intelligence" is who you truly are. The "true you" is pure Consciousness. However, you cannot know who you truly are through your opinions, stories, or by talking or thinking about it. It is a feeling; it must be felt. It lies beyond the thinking mind.

You are not your car. You are not the things that you have in your life. You are not even the ideas that you have of yourself. All of the things or ideas that you believe you are are simply labels and role playing. You will hear people say, "I am a mother of three children," "I am a teacher," "I am a doctor," "I am a cancer survivor," "I am Kathy," "I am John," etc. The question is who were you before you had children, before you became a spouse, before you survived that cancer, before you became a teacher or doctor, before your parents decided to call you Kathy or John? Didn't you exist before all of these labels? Will you not still be here even if all the labels, the titles, the careers, the degrees, the qualifications, the spouse, the children, the cars, the house, or even your name are removed? The truth is that, as humans, we play roles in life, and we become so attached to these roles and identify ourselves with the roles, the labels, the noise, and the thoughts in the head and come to believe that these labels are who we are. We believe our thoughts to be true, and we create a whole world based on these beliefs, these ideas, and these labels, which is very limiting and mostly false. The truth is you are energy that is free and liberated from ideas, labels, careers, family roles, accomplishments, stories, names, and so on. You are Consciousness. As stated by Albert Einstein, energy cannot be destroyed. Therefore, as the energy field goes, you are not created nor destroyed; you are limitless, and you are a vibration. You existed before, you exist right now, and you will be. It lies beyond the mind, and it can only be experienced by stepping into timelessness through silence.

Are you still thinking that you are flesh and bone? If yes, know that nowadays, with the advancements in technology and medicine, surgeons are able to transplant several organs such as the heart, lungs, kidney, liver, intestine, middle ears, corneas, and so on. This demonstrates that who you truly are is not the flesh and the bones because several organs of your body can be removed or replaced by synthetic materials and yet you will still be there. Furthermore, if some of your organs malfunction or are defective, they can be replaced by someone else's organs, and you will still be here; you will still be "you" without your heart, liver, lungs and so on. This shows that you are not your body. Currently, scientists and medical professionals are only able to do these transplants one or a few organs at a time, but who knows what they will be able to do in the future? In truth, you are not the body; you are not even male or female; you are pure Consciousness that has no name and that is not limited to the body.

Oneness Principles

The truth is we are energy coming from that which we call Light, Source, Universe, God, The All, Abba, Divine, The Most High, Father-Mother-God, Dadda, Mawu, Allah, Yahweh, Force of Life, All-That-Is-All, etc. However, before we can even comprehend that truth, we must first understand the truth about who or what we call "God" and the principles of Oneness.

"God" is just a word or a label that humans have invented to designate something that is beyond their understanding, which they know is there but cannot fully comprehend. As you will notice, this name "God" changes with different cultures, languages, religions, and is even given different meanings and attributes, depending on the society and the civilization, which means that it is just a label invented by humans.

In reality, "God" is just a word that is representative of an energy field, and in science, we can see the energy field, the magical light, and the spectrum that pervades all life and is around all of us. The energy is that of the Divine mind, and humans did not create the Divine mind. The truth about who we are is something that I myself sometimes find challenging to fully grasp and comprehend. Therefore, I will illustrate these concepts with a metaphoric example of the "cells" and the "body" to explain who we truly are, who or what "God" is, and the principle of Oneness. In this metaphor, what we call "God" is the body, and "we" are the cells of the body.

Oneness Principle 1: We Are All One

Cells of the body are connected to each other; collectively, they form one system, which is the body. Together, cells group to form organs such as muscle, liver, heart, brain, lung, and so on. The organs in their turn group to form individualized and specialized systems such as the cardiovascular system (heart, vessels, blood), endocrine system, nervous system, digestive system and so on. In turn, these various systems of the body also group together to form a single system that is the body. In our metaphor, the cells are individual people going about their lives; the organs are the different races or countries; the systems are different continents; the body is a single Source that we all come from and that we refer to as "God." Note here that this is a metaphor because, in truth, God is All-in-All, and God is even in the trees, the birds, the stars of the sky, the other planets such as Sirius, the air, etc. All is God, and God is All. Therefore, even though we seem to be separated and different from each other, in truth, we are all One because we all come from the same Source, the same body of God, and we form one system that we refer to as God.

Oneness Principle 2: We Are All the Same

Cells of the same body have the same genetic makeup and DNA, no matter how different and individualized they are. They are also made of the same chemical constituents such as carbon, hydrogen, oxygen, nitrogen, etc. In that manner, we can say cells of the same organism are all the same because they have the same genetic makeup and constitutive materials. In fact, from a single cell of an organism and person, scientists are now able to completely clone that organism, whether the cells are taken from the hair, the hand, or the saliva of that organism or animal. Likewise, because we all come from the same Source or Organism that is God and made of the same materials as God, we are therefore the same. We have the same propensities as God. This is what some spiritual scriptures refer to when they say, "You are made in the image and likeness of God," or "Ye are God."

Oneness Principle 3: We Are Unique and yet Equal

Even though cells that come from the same organism are the same, they are very unique. In scientific terms, this is referred to as cell differentiation. At the beginning, cells of the body are the same, but later, they differentiate to become either cardiac cells of the heart, neurons of the brain, myocytes of the muscle, and so on. Cells are very specialized, unique, and yet indispensable to each other because they have unique specialties and no cell's specialty is better or worse than another's. In fact, a body has trillions of cells. There are different kinds of cells in the body, and each cell has their unique specialties and functions. There are cardiac cells, which are specialized in pumping and distributing blood to the other cells of the body. Without the cardiac cells, the other cells in the body would not be supplied with the nutrients and minerals contained in the blood and would die. There are brain cells called neurons; neurons are specialized in receiving information and stimuli from the outside, transforming this information into electrical signals, and transmitting these signals to the other cells in the body for appropriate responses. Without the neuron cells, the other cells would be without direction, the body would be without a brain, and the person would not be able to think, imagine, create, or even exist. There are lung cells, which enable the body to breathe and take in oxygen from the air while helping the body to get rid of carbon dioxide. Without the lung cells, the other cells would be depleted of oxygen and would die. We can see that no cell is below or inferior to others because they are all needed, and the role of one cell is as important as the role of every other cell in the body. Therefore, we can say that cells are all equal in their importance, roles, and worth.

Similarly, even though we, human beings, are the same and come from God, we are also unique and inimitable. You are sacred, and there is no one else like you on the earth plane—never has been, and never will be. Nobody is below or above others. No race is superior or inferior to others. No gender is better or worse than the other. No role is lower or higher than the other. We are all equally important, equally valuable, and equal in worth because we are simply individualized or differentiated expressions of the same Source or Power that we call God. Everyone is a wonderful, sacred, and unique expression of God.

Oneness Principle 4: We Have the Same Potentiality

Cells of the body have the same genetic makeup, constituents, and potential. In fact, scientists have discovered that before cells differentiate, they are in a form called "stem cells". When a stem cell is removed from its original organ or location in the body and placed in another environment in the body, it can differentiate and perform at the same level as the other cells in its new environment. This is the basis of cloning and other new scientific advancements nowadays. Likewise, we, human beings, have the same potential, no matter who we are. Whether we are male or female, young or old, rich or poor, we all have the same potentiality and have the ability to do and create amazing things. In fact, hundreds of years ago, people thought that women could not accomplish certain tasks or even go to school. Today, it is obvious that women have the same potential as men. Centuries ago, it was believed that some races or ethnicities are limited or cannot accomplish certain things, but today, there are demonstrations everywhere that this is not true, as we have people of all races and ethnicities accomplishing wonderful deeds and creating amazing things in this world. The truth is that we all have the same potential as our Source God, and we are all able to go as far as any other person.

Oneness Principle 5: We Are All Connected

Cells of the same body are connected to each other. For instance, cardiac cells are connected to the muscle cells of the toes, even though they are located on different parts of the body and separated from each other. In fact, if something happens to cardiac cells, and there are imbalances, and they become ill and malfunction, this illness of cardiac cells will also affect the cells of the toes, as they will no longer be supplied with enough blood, and there may be an imbalance in the toe cells as well. Indeed, cells of the body are all connected and respond to each other through what biologists call feedback loop systems, which are biological mechanisms that allow the body to maintain homeostasis (stable equilibrium and balance within the body). There are positive and negative feedback loops. For instance, if for some reason the body temperature increases, a feedback system is activated to initiate cooling mechanisms such as sweating, and the body temperature will decrease—this is an example of a positive feedback loop. All the cells of the body are connected to each other and respond to each other via these feedback loop systems. Anything that happens to a cell of the body will affect the other cells of the body, which will respond via the feedback loop systems. For instance, if the lung cells develop malignancies or cancer, sooner or later this cancer will affect all the other cells of the body, and the whole body will be ill.

Similarly, we are all connected to each other on Earth, whether we live a few blocks away or thousands of miles away from each other. Anything that happens to someone on this earth plane affects the rest of humanity, whether it is positive or negative. The spillover or feedback loop mechanisms are always in operation whether we are aware of it or not. For instance, a country cannot create a little mess in some parts of the earth by bombarding other people without affecting the rest of humanity and its own people by paying the price of lives lost and military families destroyed in its own population. Anything that anyone does will come back to them like a boomerang, slapping them in the face. In the same way, any positive action that you put forward—such as helping others, praying for others, supporting the poor in your community, or even the smallest act of kindness such as smiling at a stranger or opening the door for others—will affect the rest of the world in some way and will come back to you as a blessing via the positive feedback loop. The truth is that we are literally connected to each other by the combs of an invisible net.

Oneness Principle 6: We Are All Part of a Greater Life Even Though We Have Our Own Individualized and Complex Lives

The cells of the body have a world of their own. Cells create things, synthesize proteins, and are very organized among themselves to work efficiently. Cells multiply by the mechanism of mitosis, and they die by a process called apoptosis. A cell has a whole life of its own—from birth to work life to death. A cell is very organized and intelligent, and it has the ability to analyze its external environment and to make a decision on the best action to take, etc. The process of synthesis of protein using the genetic code is so complex, so well organized, and so dynamic that it is undeniable that a cell has a life of its own and lives a life full of complexities, challenges, and work. Cells even get sick and sometimes find ways to heal themselves or to cope with the sickness or the imbalance. There is a whole life inside a single cell. However, even though cells have their own lives, they are at the same time part of a greater life, which is the body, and they belong to something bigger than themselves. To the cells, the body is like God.

Likewise, humans have their own individualized lives. We are born, go to work, and we create even complex materials such as robots, airplanes, and computers. Humans are very intelligent; we mate and give birth to other humans that we call our children. As humans, we have challenges; we live and then pass away via a process called death. However, we, humans, are also part of Greater Life. To us, we call that Greater Life "God", just as the body is like God to its trillions of cells.

Illustrative Analogies and Metaphors of "Who We Are"

We are energy. We are not our bodies. We are souls coming from God. We are made of the same substance and essence as God. Here are some analogies to illustrate this:

1. God can be seen as a tree, and we all are as leaves on this tree and all feeding from the same tree. That is why we can seem to be separate beings, but, in truth, we are all connected, and we are all One. The whole Universe can be seen as God, and God constitutes the whole Universe. This means that God is everywhere, omnipresent. God is in the tree, in the air, and in everyone.

2. We can see God as the ocean, and each one of us is like a drop of water that makes up this ocean. Notice that if you take a drop of water from the ocean, this drop has the same constituents and

chemical properties as the ocean. This single drop has the same salt and the same constitutive chemicals as the original ocean from which it comes. The same thing applies to us as well. We have the same properties and propensities as God.

3. We can see God as someone who created a video game (Universe) and at the same time is playing the various roles of that video game. This means everybody is just an extension of the same player, God, who is playing different roles of the game of Life.

You may ask why did God create the Universe and what is the purpose of all of this? The purpose of creation is for evolution, fun, challenges, curiosity, experimenting new things, and the joy of creating and playing.

Why Are We Here on Earth?

The earth is a school. It is a third-dimensional-plane school where we come to learn how to love, grow, evolve, create, and enjoy the process along the way. There are billions and billions of galaxies, and each galaxy has billions of planets and systems in it. This is more than our human brain can comprehend. "Infinity" is the best word to describe the immensity and magnificence of the Universe and what we are part of. We come to Earth to learn, create, and love. We all come here of our own accord and out of love. No one forced us to be here; we came because we decided to come to learn, create, and evolve. We purposely signed up for the schoolroom of Earth, knowing what to expect.

Everybody has a series of lessons that they chose prior to their birth, and everybody has a life purpose that is uniquely theirs. However, the ultimate lesson or purpose is to learn to love: to love everybody everywhere no matter who they are.

THE TRUTH ABOUT LIFE

After a series of incarnations, several thousands of incarnations or whatever number of incarnations is needed, we graduate and continue on another dimension, a higher dimension such as the fourth, fifth, or sixth dimension, where there will be less pain, less suffering, and more love. Third-dimensional planes such as the earth plane are not easy, and the earth is one of the challenging ones. It takes courage to sign up for the schoolroom of Earth, so you can be proud of yourself for your courage and willingness to be here. Here on Earth, we learn how to choose love over fear. The earth is also called the great school of duality, where we come to experience opposites: male and female, day and night, tall and short, love and fear. However, in truth we are made of Love, and there is only Love in heaven where we come from. That is why it seems really difficult for us here because Earth is not our natural home. Some complete their lessons on the third dimension quickly after a few thousand incarnations, while others take eons to learn and complete their lessons. The ultimate goal is to awaken of the illusion of fear and to become unconditionally loving, which we all are capable of in spiritual truth. The good news is that there is no race, as there is nothing to compete against because we have eternity. Therefore, we can come as many times as we need or want to the schoolroom of Earth. In fact, this is good news because it means that we have endless chances and opportunities and that we can try as many times as we want. However, every time we come back, we forget everything and start anew. This is the rule of the game. I believe it is set up this way so that we can experience everything to the full, from birth to death. It also makes the game of life more interesting and realistic. The goal is to wake up from the illusion and realize the truth about life and about who we truly are: extensions and expressions of God.

Where Do We Come From and What Are We Doing Here?

We come from the various parts of the Universe to learn, grow, create, and evolve spiritually. We come from diverse planets and diverse galaxies to attend this unique school of Earth. In spiritual truth, only a part of our energy or soul comes here to learn. Our Higher Selves remain in heaven. We can also have other parts of our energies or extensions of our souls in other planets or galaxies, learning other things at the same time. All is possible. In a sense, we are omnipresent like our Father God. The earth is a particular school because of the diversity of souls and races that it has. It makes this school more interesting and more challenging at the same time, as it gives us the opportunity to learn to love those who are different from us: different races, different religions/beliefs, different ways of living, and so on. The earth should not be seen as a "melting pot" because of the variety of souls and experiences that it provides; instead it should be seen as a "salad bowl" where various souls make up the different ingredients, each uniquely contributing to making the bowl of salad. A salad will not be tasty without the lettuce, the dressing, the tomatoes, the onion, the croutons, or the avocado. Indeed, it takes all the ingredients mixed up to make up the unique flavor and taste of salad. Similarly, the salad of Earth will not be tasty without all the diversities of the ingredients in it. Indeed, it takes every race, religion, sexual tendency, culture, belief, everything to make up this unique, unprecedented, and amazing school of Earth.

There are also different beings of different dimensions on the earth right now. Even though the earth is a third-dimension planet, there are beings of fourth, fifth, sixth, seventh dimensions, etc., among us right now. These souls came to help people on Earth, as the planet itself is moving to the fourth and fifth dimension. These souls are what some refer to as "light workers". There are also incarnated angels among us;

some refer to them as "Earth angels". However, it does not matter who came from where because we are all in the same boat, and we all experiencing the same challenges of living in a third-dimensional plane. The truth is, we all come from somewhere. If we consider the earth to be the third grade, for instance, no one truly starts school at third grade but kindergarten or preschool. So, with that in mind, it means that we all come from somewhere to attend this great school of duality that we call Earth.

THE PHYSICAL BODY

"The truth is incontrovertible. Malice may attack it, ignorance may deride it, but in the end, there it is."

–Winston S. Churchill

What Can You Tell Us about the Physical Body?

The Body Is Pure Energy

The body is made of energy; everything in the body is made of energy. This is one of the grand illusions that many do not understand because the physical body appears to be solid to the touch; thus, they believe that the body is just a pure mass of flesh, bones, and blood. However, scientific researchers have shown that the body is nothing but pure energy. Indeed, the great scientist Albert Einstein demonstrated this with his famous formula **E=mc²** (where **E** represents energy, **m** is the mass, and **c** is the speed of light). With this mathematical equation, we can deduce that mass (matter, body) equals energy divided by the square of the constant **c**. Therefore, the mass (matter, body) is energy since **c** is just an invariable constant.

$$m = E\backslash c^2.$$

In our example, the mass **m** represents the body. This shows that the body is just pure energy. You may ask how this can be possible because when you look at your body, you see only solid matter and not energy. Nowadays, with the help of the science of biology, chemistry, quantum physics, and the advanced technologies that we have such as microscopes, thermal pictures, frequency and vibration measurement instruments, we know that the body is made of energy or, to be more precise, the body itself is just pure energy.

The following illustration is a demonstration or a zoom-down to show what the body is made up of:

The body is made up of different organs such as heart, bones, lung, etc. Each organ is made up of tissues such as muscle tissue, bone tissue, etc. Each tissue is made up of cells such as neurons or brain cells, nephrons or renal cells, red blood cells, etc. Each cell is made up of molecules such as proteins, carbohydrates, lipids, etc. Each molecule is made up of atoms such as carbon, hydrogen, nitrogen, etc. Each atom is made up of subatomic particles such as protons, neutrons, electrons. Each subatomic particle is made up of quarks. Quarks are energy with angular momentum, linear momentum, spin, electric charge, and color (quantum chromodynamics). When we go further, we will see that quarks are made of oscillating or vibrating waves of energies. Therefore, we can conclude that the body is made up of energy, or, more correctly, the body is energy.

The Body Is Pure Energy

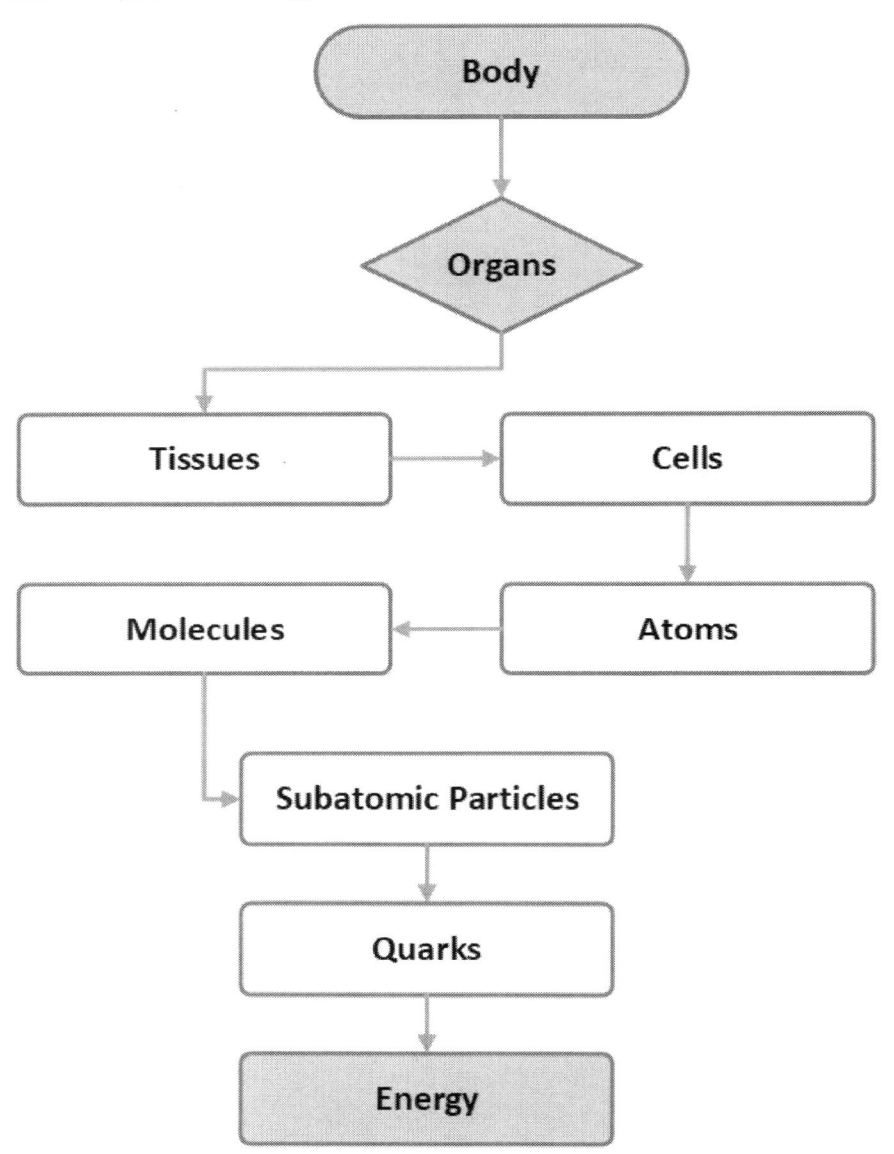

```
Body
 ↓
Organs
 ↓
Tissues → Cells
           ↓
Molecules ← Atoms
 ↓
Subatomic Particles
 ↓
Quarks
 ↓
Energy
```

An analogy to illustrate this is perhaps the case of water. Water can exist in various forms such as liquid, vapor, and solid (ice). When you put a bowl of water on a stove, for instance, and the water is boiling and evaporating, it dissipates in the room, filling the room with vapor. However, you may not see the vapor with your physical eyes once it is dissipated, but that does not mean there is no water in the room; there is water in the room but just in a form that you are not able to see with your physical eyes. This is the same case for the body as well. The body is just a condensed form of energy, and just because you cannot see its spectrums of energy with your eyes does not mean that the body is not energy.

The Body Is Made Up of Light and Contains Various Energetic Centers or Systems

According to the great physicist Albert Einstein, energy and matter are the same, and all matter is comprised of energy. This means that the body (matter) is made up of energy. We know that cells are made up of atoms, which, in turn, are made up of subatomic particles; these subatomic particles are made up of quanta of energy waves photons and are filled with light. Therefore, we can conclude that, to its very core, the body is made up of light.

The body is not only pure energy and light, but it is also an amazing, complex, extraordinary system that contains layers of energetic fields, which intertwine, superpose, and interconnect in a sophisticated, scientific, and magnificent manner. We can, therefore, say that the body is an ensemble of energetic system fields of various light spectrums. Indeed, inside the body, there are energetic centers whose roles are to nourish and replenish the energetic systems of the body. These energetic centers, called chakras, function somehow like little batteries to fuel and replenish the body.

There are seven chakras located inside the body, but, in truth, there are approximately 12 chakras, including the seven that are located inside the physical body and the remaining five located above and below the body. Scientists and old civilizations, as well as the Eastern tradition, are well aware of the existence of these energetic centers. These chakras are often represented with various colors, and, in truth, the body and its energetic system are made up of light and various colors as well.

Besides the chakras, there are meridian systems inside the body that work as entry energetic points, which are scientifically proven, well known in the Eastern traditions, and are the basis of acupuncture modalities. Unfortunately, even today, many humans still don't know that they are an energy field, and they do not know what it looks like.

What makes us believe that the body is just a solid mass and not an energetic field is our inability to perceive energy, to see the light spectrums that make up the body, or the colors of these energetic centers. In truth, there are various spectrums of light. Each type of light has a unique wavelength, but the human eye is only able to see a very tiny portion of the various spectrums of light that exist. The great majority of spectrums of light cannot be seen with the naked eye. In fact, the visible spectrum of light that can be seen with the eyes is a very small segment of electromagnetic light that ranges approximately between 400 and 750 nanometers. Anything below or above this small range of frequency of light cannot be seen with the physical eye, but that does not mean that they are not there.

Visible Light Spectrum

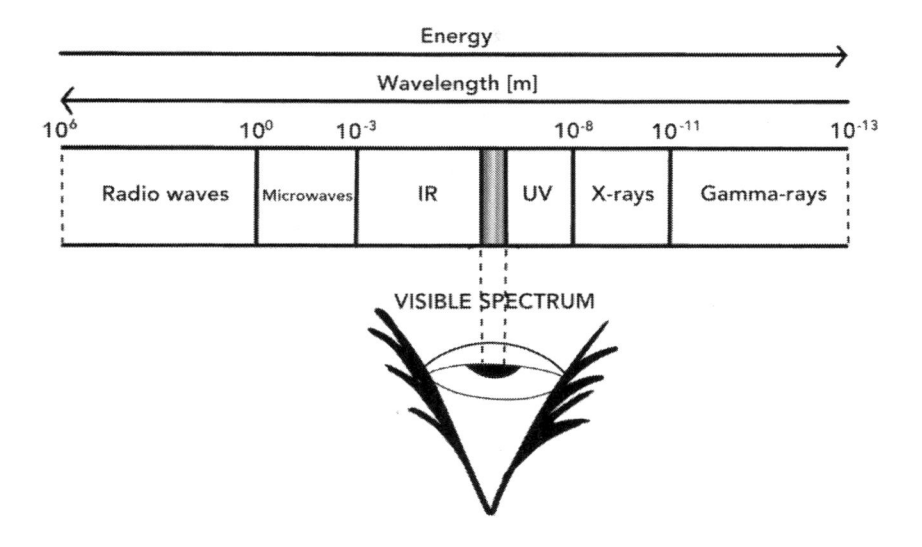

```
                        Energy
        ←————————————————————————————————→
                    Wavelength [m]
   10⁶        10⁰      10⁻³      10⁻⁸    10⁻¹¹      10⁻¹³
   | Radio waves | Microwaves | IR | UV | X-rays | Gamma-rays |
                              VISIBLE SPECTRUM
```

You Are Not the Body but the Consciousness That Inhabits and Moves Your Body

This Consciousness itself is pure energy, pure Light. Some refer to it as soul, Essence, Spirit, and so on. Who you truly are is that Consciousness, but it is very tricky because you cannot see your Consciousness; thus, you may think that you are just your body, but you are not. You are not even a body that has Consciousness but a Consciousness that chooses to wear a body like a costume. This is the great masquerade of life, the game of life, the great theater where the truth is hidden, yet the truth is still right in front of our noses for those who look a little deeper.

The body seems to exist in space and seems to stop at the surface of your skin, but, again, you are not your body; in fact, you are rather the Consciousness that is wearing this costume that you call your body. Your Consciousness expands beyond your body. Just like when you wear a T-shirt and a pair of pants, your head, legs, and arms expand beyond your T-shirt and pants; in the same way your Consciousness goes beyond your body. The energy field that you truly are expands beyond your physical body. You can see yourself as an eggshell of energy field and to enable this eggshell of energy field that you are to move from place to place, there must be a vehicle.

Your body is a vessel or the physical vehicle that enables you to travel around this planet and to have a physical experience. If your neighbor is driving a Toyota Corolla, for instance, you will not say that your neighbor is his Toyota Corolla car; in the same way, you should not confuse who you truly are (Consciousness) with your physical vehicle, which is your body. The body is a vehicle, and just as a car cannot move itself without a driver, your body cannot move by itself without your Consciousness, which inhabits it. You will not bring your car, for instance, to your dentist to fix if your car has a problem; likewise, you must know how the body works and what it is made up of before even attempting to fix it in times of imbalance or disease.

The truth is that, just as you have a physical body, which is a condensed and solid form of energy, you also have other energetic bodies that are not condensed like your physical body but instead are fluid, pure light, and invisible to the physical eyes. In fact, you also have a spiritual body, an emotional body, and a mental body. To understand this, you can see yourself as an eggshell of energy field, with several layers of various energies overlapping around each other.

For instance, if you imagine yourself as the flame of a candle, your various energetic fields or energy bodies (spiritual body, emotional body, and mental body) will be like the different zones or parts of this candlelight. For example, with a candle flame, there is a blue outer edge of the flame, a yellow central zone of the flame, an orange\brown inner part of the flame, and so on. All these various zones of the candlelight are of different temperatures, some in the hundreds of degrees Fahrenheit and others in the thousands; yet, these different zones of the candle flames overlap each other to form a unified energetic field that we call candlelight. However, when you see candlelight, you will just see a light because your eyes may not be able to see that it has various layers of different colors.

It's the same thing with a human being as well. A human being is made of several layers of energetic fields (energy bodies) that serve different purposes and have various electromagnetic properties. Yet, some of these energetic fields such as the physical body can be seen, while others such as the spiritual body, the emotional body, and the mental body cannot be seen by naked eye. In truth, we are amazing energy beings, light beings, made of various beautiful and colorful spectra of light that are far beyond our human comprehension and current scientific knowledge.

<u>Various Parts of Candlelight</u>

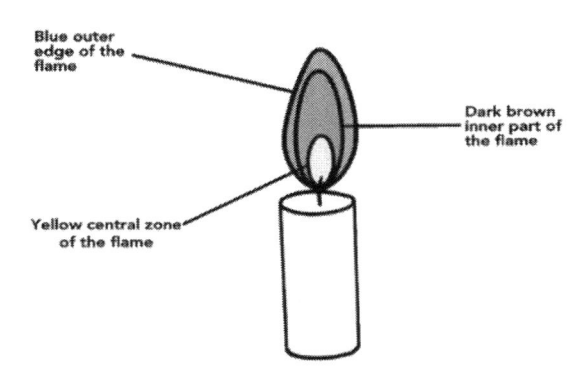

You are pure Consciousness, but this is tricky because when you look at yourself, you just see your body and not your Consciousness; thus, you believe that you are your body. For instance, let's suppose that a child is looking at a tree from the window of their home and sees the branches and the leaves of the tree swinging from side to side. The child may suppose that the tree is moving by itself with his or her limited understanding. However, what is happening in truth is that there is a wind that is gently blowing and moving the tree, which of course the child cannot see because it is invisible to the naked eye. Similarly, it is your Consciousness that animates, gives life to, and moves the body. The body is just the vessel, the vehicle, the temple of your Consciousness/soul. Your Consciousness existed before the body and will exist after the body; it is eternal. For proof, when the Consciousness leaves the body, the body becomes inert, un-alive, and people call this death. Some have reported seeing a white light or an energy leaving the bodies of their loved ones at the time of death, and they are right. The Consciousness is pure light, pure energy, and pure Essence.

The Chakra Systems

You Mentioned the Existence of an Energetic System within the Body Called Chakras. Can You Tell Us a Little More about Them?

Chakras are energetic centers that function like energetic batteries to energize, replenish, fuel, and support the physical and emotional body. They can be seen as small disks of energetic batteries that turn round and round to fuel the body's energy systems. When these chakras are opened and function well, the energy systems are balanced, and the body is healthy. However, when these chakras are closed up or not fully functional, which can happen if there is emotional suffering and pain, stress, and a fearful mind, the energetic systems of the body may become unbalanced, and, in the long term, the body may develop diseases if these issues are not addressed and released.

Emotional pain, stress, suffering, and fears are forms of energy as well. According to the first thermodynamic law, also known as the Law of

Conservation of Energy, "Energy can neither be created nor destroyed; it can only be changed from one form or another" (Albert Einstein). Therefore, if these negative energies are not addressed or transmuted, they will create disturbances in the normal flow of the energetic system of the body, and, with time, these negative energies may be changed into other forms of energy, more condensed forms of energy, which will show up in the body later as imbalances or diseases.

What many people do not know is that the negative energies they hold in the mind and toxic feelings can change form, crystallize, and show up later as disease, just as the vapor of water can change from the state of vapor to liquid-water and then to ice. It is all science. I will expand more on this in the subsequent chapters of this book.

The chakras can also be seen as energetic pumps, energetic regulating centers. There are seven chakras located within the physical body and five others located above or below the physical body. There are 12 of them in truth. However, for practicality, we are going to focus only on the seven chakras that are located within the body. When these chakras are opened and fluidly flowing, you experience health, well-being, and equilibrium on all levels.

Root chakra:

This is the first chakra and it is located at the base of the spine. Its associated color is rugby red. Its function is related to grounding on the earthly life, security, the feeling of belonging to the planet, receiving abundance, and safety. Therefore, if you are having fearful thoughts about your safety, or you are stressed out about financial security, or concerned about your survival for tomorrow, these may lead to energetic blocks in your root chakra and may later lead to imbalances or diseases of organs associated with this chakra.

Sacral chakra:

It is located just below the navel or belly button and associated with the color orange. It is the seat of creativity and passion. Any issue or fear connected to creativity or passion will lead to disturbance in this chakra center.

Solar Plexus:

This is the third chakra. It is located in the stomach area, and its related color is the golden yellow color of the sun. It is the seat of willpower, control, self-expression, confidence, and attracting positive intentions from others. Power struggle issues, control issues, arrogance, selfishness, pride, feelings of being small or unlovable, or feelings of not being good enough will manifest later as blockades in this beautiful chakra center and will impede the energy flow in that location of the body, which, if not addressed or released, will lead to diseases of the glands and organs located in this region of the body.

Heart chakra:

Located in the heart region, its color is emerald green. It is the seat of unconditional love, self-love, loving others, compassion, empathy, and the ability to see and feel connected to others and Oneness with all people. Heartbreaks, disappointments in relationships, closing your heart to loving others, etc. will close the heart chakra center and may later lead to heart diseases.

Throat chakra:

Located in the throat area, its color is blue, sapphire blue. It is associated with speaking your truth with integrity, living a truthful life, creativity, and expression of talents and gifts. Lying to yourself or to others, not speaking your truth for fear of judgments from others, being dishonest in your speech and actions, deceitful speeches or written communications either in your personal life or in your work life such as being a salesman and telling small lies to clients, and anything related to communication issues will lead to blockages in the throat center and may later manifest as imbalances and diseases in this area of the body.

Third Eye:

This chakra is located in the area between the two eyes, and it is the seat of intuition, seeing the divine truth. Not seeing the truth, seeing separation such as believing that others are separated above or below you, or believing that you are merely a body or an isolated human struggling alone on Earth may lead to energy blocks in the third eye along with its related health issues.

Crown chakra:

Situated on the top of the head, its color is white or royal purple. It is associated with the connection to the Divine and spirituality. Not realizing your divine nature, having no connection to the Divine Mind that we call God, or to your angels, or your guides, feeling deep despair and angst against God, not believing in anything, etc. may create blockades in the crown chakra center and may lead to imbalances and diseases later.

Just as the heart pumps, fuels, and distributes blood and nutrients to other organs of the body, and when there is a blockade in the heart, and the blood is no longer being supplied to the organs, it will lead to disease in the body, similarly, when there are blocks in these energetic chakra centers that fuel and supply energies to the body and its organs, imbalances and diseases will manifest in the long run. Inversely, when these chakras are fully functional and open, the body's energy fields are well balanced, energized, and healthy.

The Truth about Life
CHAPTER 3

HEALTH AND DISEASE

"A lie is like a snowball, it starts off small and then grows and grows until a point where it gets so big it falls apart and then the truth is discovered."

–Chris Hughes

Why Do We Get Sick, and Where Do Diseases Come From?

It is undeniable that thoughts are able to create physiological reactions and biological changes in the body. For instance, a thought about an ice cream or a favorite dessert can make you salivate, and a thought about a sexual fantasy can produce changes in a man's sexual anatomy. The truth is that every biological change—from the slightest, such as salivation, to the biggest, such as diseases—has at its core origins electromagnetic aspects that can be retraced back to thoughts, feelings, and emotions. Indeed, thoughts initiate and trigger a series of cascades of reactions in the body that lead to complex biochemical changes, thus affecting the body's cells structures and organ functions.

This knowledge seems to go against the current understanding of science, which tends to emphasize that everything is coded in the genes. However, new evolving fields of science such as epigenetic, psychoneuroimmunology (PNI), quantum physics, etc., are showing that the old dogma of the scientific doctrine solely focused on genetics may not be 100% accurate.

For instance, in the Adverse Childhood Experiences Study (ACE Study) conducted by Kaiser Permanente Hospital in partnership with the Center of Disease Control (CDC) involving over 17,000 patients, the relationship between Adverse Childhood Experiences (ACE) and various health outcomes was analyzed. In this large-scale study, certain childhood stressors such as having alcoholic parents, parental separations, having one parent on substance abuse, domestic abuse, and mental illness of a parent were scored. The study showed that people raised in such families are:

- Twice as likely to be diagnosed with cancer
- Seven times more likely to become alcoholic
- Thirty times more likely to attempt suicide
- More likely to suffer chronic health conditions such as diabetes, stroke, lung diseases, heart diseases

Men who scored poorly in the ACE study were 4600% more likely to use intravenous drugs. The study unquestionably showed that environmental factors such as having a dysfunctional familial environment are directly linked to chronic health conditions and social problems across lifespan. This study is an eye-opener and points out that the origins of health issues may be outside the genes. The questions to ask are these: Why are people raised in abusive and dysfunctional households more likely to develop serious health issues? Would these people have the same fate if they had been raised in loving households? What is it about them that makes them more likely to have chronic health diseases and to become alcoholic or IV drug users? Aren't they possibly carrying traumatic emotions from their childhoods? How did these heavy childhood emotions translate later into physical diseases? What would their lives be like if after their childhood experiences they went through years of therapy, emotional support, meditations, prayers, and on forgiveness journeys to release the pains and emotions of their childhood?

Furthermore, in his book, *The Genie in Your Genes*, Dr. Dawson Church points out that genes are affected by external environmental factors such as thoughts, emotions, feelings, and intentions. He demonstrated that these external factors are able to turn on or turn off gene expressions, therefore inducing a series of biological effects in a dynamic complex dance. Indeed, Dr. Dawson showed in his book that some genes called "immediate early genes" are able to react to environmental circumstances within minutes. Even though they have their peaks in approximately an hour, many of these immediate early genes react to changes in the environment such as stress or emotional distress very swiftly and can be expressed quickly—within seconds or in less than two minutes. Among these immediate early genes is a family of gene called c-fos, which reacts to stressful situations such as situations involving immediate danger or being in heated arguments. According to scientific findings, these stressful situations trigger the brain to produce a protein called fos. Fos then binds to the DNA molecules, initiating the transcription of other genes, which in turn leads to a cascade of biochemical reactions in the body in order to deal with the stressful situations. This perfectly shows that external factors such as the negative thoughts and feelings engendered by being involved in heated arguments or being in a dangerous situation can change the molecular expression of genes in the body and alter cell functions.

Currently, many people believe that diseases just happen and that the biological mechanisms of their bodies one day just start to malfunction out of nowhere. It often goes like this: People wake one day and they discover that they are not feeling well. They go to their doctors; their doctors give them a diagnosis, and they are told that they have such and such disease. Indeed, with current scientific knowledge and understanding, the causes of diseases are explained by cellular malfunctions or biological mechanism abnormalities, or sometimes the causes of disease are simply unknown. In other words, the causes of the diseases are confined or limited to biological elements such as genes, cells, molecular processes, protein disruptions, organ system malfunction, and so on. However, the questions that remain are: what causes the body's cells to develop these abnormalities or molecular irregularities to begin with? Do the cells just snap one day out of nowhere and start to develop issues and diseases? The answer is NO.

The truth is that diseases are caused by subtler vibration elements such as thoughts, feelings, and emotions that cannot be perceived with the naked eye or under the microscope. This can seem to go against our current medical knowledge. However, keep in mind that some scientific breakthroughs such as penicillin, aseptic hand washing, balloon angioplasty, and aspirin were discovered decades or centuries before the scientific and the medical world proved their effectiveness or sometimes were simply rejected before they become accepted procedures or commonly used drugs.

In truth, diseases are caused by repetitive negative thought patterns, emotionality, and spiritual disconnection, which are then expressed as imbalances on the physical level in the body as diseases. Indeed, thoughts are not just thoughts. Each thought has electrical frequency. Scientists have discovered that the brain produces electrical signals and the nerves in the brain fire up as electricity. This discovery is the basis of electroencephalogram (EEC). However, people do not know that each thought carries a specific electrical frequency. Positive thoughts such as loving, kind, and joyful thoughts have high vibration frequencies. On the other hand, negative thoughts such as angry, hateful, jealous, and resentful thoughts have low vibration frequencies that are detrimental to cellular activities and ultimately lead to disease.

For instance, Electroconvulsive Therapy (ECT) is a new procedure for treatment-resistant chronic depressions in which small electric currents are passed through the brain. Although healthcare professionals do not know how ECT exactly works, they know that the procedure seems to cause changes in brain chemistry, which help to reverse the symptoms of major depression. This supports the idea of the possibility of involvement of electromagnetic components in some diseases and clearly shows that small electric currents do create changes in the chemistry and molecular constituents of the brain. In fact, thoughts carry electric frequencies that affect the cells in the body and trigger cellular synthesis of biochemicals, hormones, and neurotransmitters, which in turn affect organs and the mechanisms of the body, leading to biological changes in the body such as imbalances and diseases (in the case of negative thought patterns) or health and restoration (in the case of positive high-frequency thought patterns).

To understand this, you must be able to comprehend the body and its energetic systems and the cellular phenomenon that the body is in truth. For instance, if you observe the body under a microscope, you will see that the body is a cluster of vibrating cells. When the human brain thinks thoughts, they are all of certain frequencies, which can even be measured. Indeed, nowadays, scientists have advanced technologies and tools to measure these frequencies. What often happens is that repeated thought frequencies impact the structure of the body. High-frequency thoughts such as loving thoughts, happy thoughts, joyful thoughts, laughing thoughts, and thoughts of gratitude promote the rejuvenation and restoration of cells and the and a harmonious functioning of the body's mechanisms. On the other hand, low-frequency thoughts such as anger, cynical thoughts, worrisome thoughts, fearful thoughts, resentful thoughts, judgmental thoughts, and so on, create cellular imbalances, mutations, cell malfunction, and diseases in the body.

This has been scientifically proven in plants. When you speak kindly to a plant and project beautiful images and thoughts to the plant, it grows; the cells of the plant are healthy, and their structures are intact. On the other hand, if you do the same thing to a different plant using negativity, destructive thoughts, and language, you will see that the cells of that second plant will crumble, malfunction, mutate, and may have diseases.

Another component that plays an important role in the manifestation and causes of disease is our feelings or emotions. We have seen that thoughts are electrical signals that fire out from the brain and affect body cell structures and chemicals. On the other hand, feelings are "magnetic signals" that create waves of vibration or frequency, which diffuse and bathe the cells and organs of the body.

Thoughts = electrical signal
Feelings = magnetic signal

Thoughts coupled with feelings create a pull and a compelling wave of electromagnetic signal. When this electromagnetic signal comes in contact with the energy field of the body, it affects and changes the natural electromagnetic field and frequencies of cells. This leads to subatomic particle conformation changes, and atoms reorganize themselves differently and are disturbed. Consequently, abnormal chemicals are produced from these deformed subatomic reorganizations. These abnormal chemicals affect cell functionalities, and the cells change or mutate. Then the organs that these cells are part of display these mutations and abnormalities. Consequently, the body displays diseases. Diseases are only the final stop of a cascade of reactions initiated by thoughts and feelings, which most of the time may be going on for months or even years before the abnormalities or imbalances surface as diseases in the body.

Thoughts and Feelings

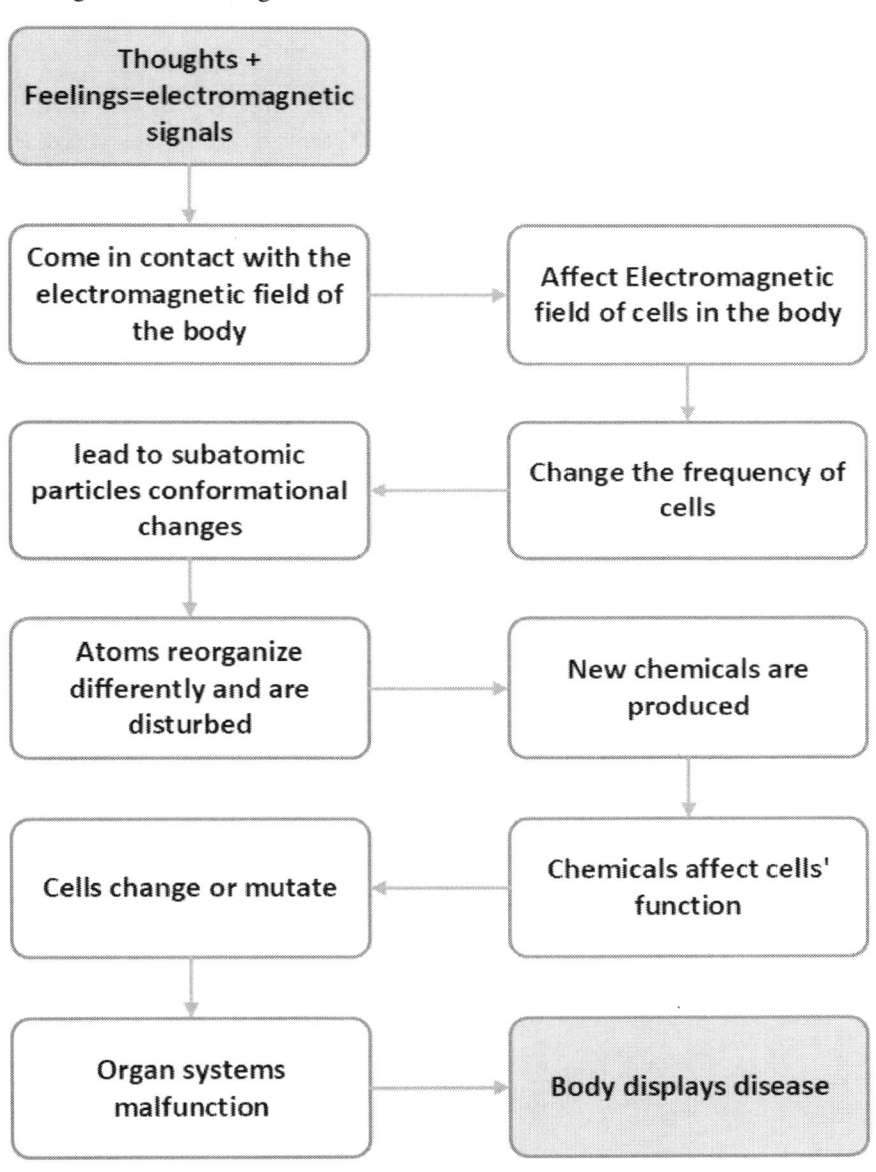

```
Thoughts + Feelings = electromagnetic signals
            │
            ▼
Come in contact with the electromagnetic field of the body  ───►  Affect Electromagnetic field of cells in the body
            │                                                                    │
            │                                                                    ▼
lead to subatomic particles conformational changes  ◄───  Change the frequency of cells
            │
            ▼
Atoms reorganize differently and are disturbed  ───►  New chemicals are produced
                                                              │
                                                              ▼
Cells change or mutate  ◄───  Chemicals affect cells' function
            │
            ▼
Organ systems malfunction  ───►  Body displays disease
```

Although many people do not know the truth about how diseases manifest in the body, an increasing number of people are becoming more aware of the truth. For instance, a recent scientific research has shown that 70 to 75 % of all health issues may be related to stress. Furthermore, many scientific studies have proven that stress increases the level of the cortisol hormone in the body. A high level of cortisol hormone, known as the stress hormone, is associated with many health issues such as blood sugar unbalance, weight gain and obesity, immune suppression, cardiovascular problems, decreased level of natural killer cells in the body, and possible early onset of Alzheimer's disease.

The question to ask is: What is stress and what causes stress? In fact, stress is caused by fearful, worrisome thoughts associated with feelings. Indeed, according to Merriam-Webster Dictionary, stress is defined as "a state of mental tension and worry caused by problems in your life, work, etc., something that causes strong feelings of worry or anxiety."

Stress is nothing but fearful thoughts, worrisome thoughts, mental pressure, nervous tension, and thoughts about what might happen in the future. Therefore, we can say that these fearful and worrisome thoughts and their effects that we label as "stress" in our society are the underlying causes of several health issues. Currently, according to our actual understanding, scientists and healthcare professionals believe that a certain percentage of disease is related to stress while other diseases are caused by spontaneous cell malfunction, spontaneous genetic mutation, environments, or simply the causes are unknown. However, the truth is that all health issues are created by negative and fearful thought patterns, unhealed resentments, accumulated anger, packed-up toxic feelings, unforgiveness, repressed emotions, and disconnection from love.

Low-frequency thoughts:

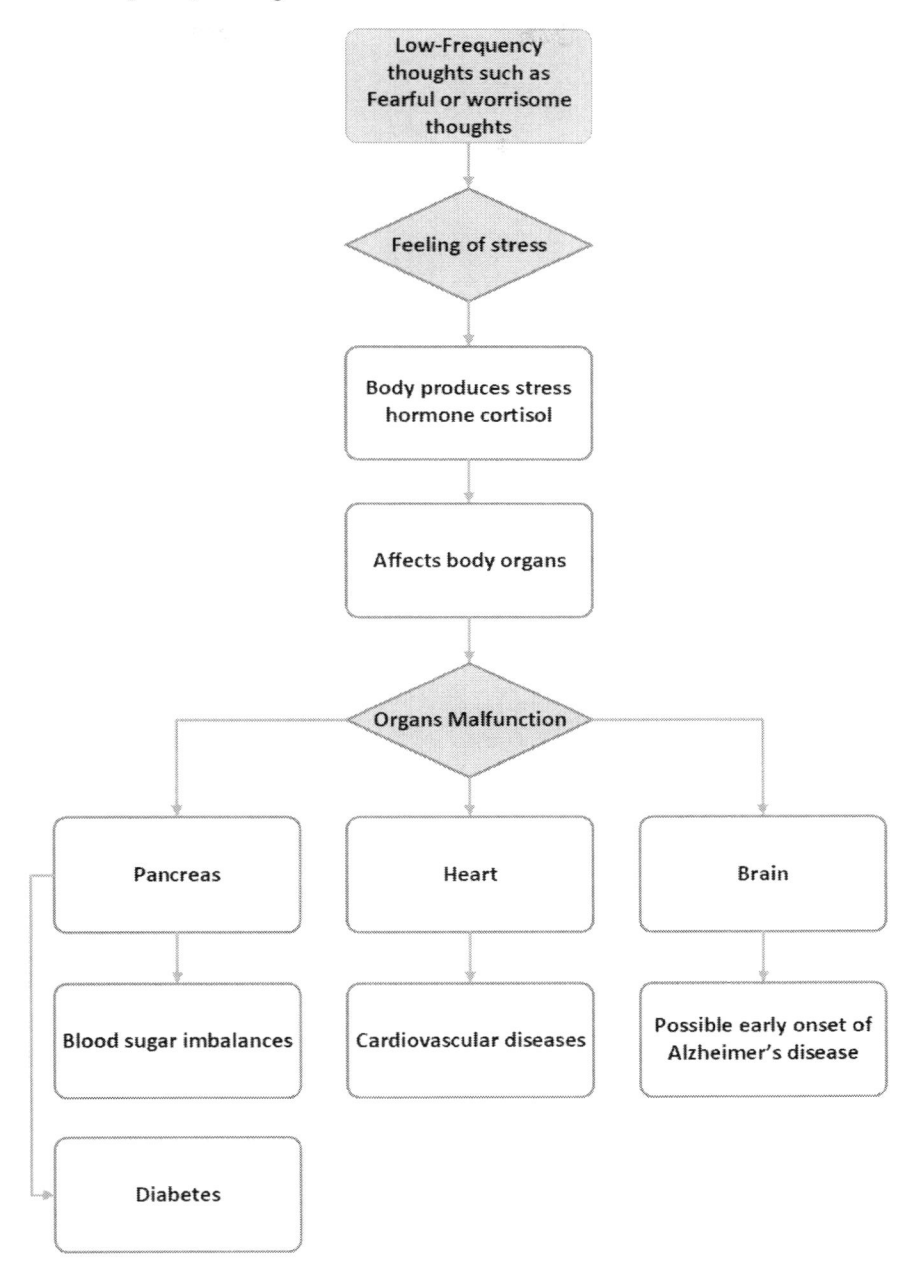

```
Low-Frequency thoughts such as
Fearful or worrisome thoughts
            │
            ▼
    ◇ Feeling of stress ◇
            │
            ▼
  Body produces stress hormone cortisol
            │
            ▼
      Affects body organs
            │
            ▼
     ◇ Organs Malfunction ◇
       │      │       │
       ▼      ▼       ▼
  Pancreas  Heart   Brain
       │      │       │
       ▼      ▼       ▼
  Blood    Cardio-   Possible early
  sugar    vascular  onset of
  imbalances diseases Alzheimer's disease
       │
       ▼
   Diabetes
```

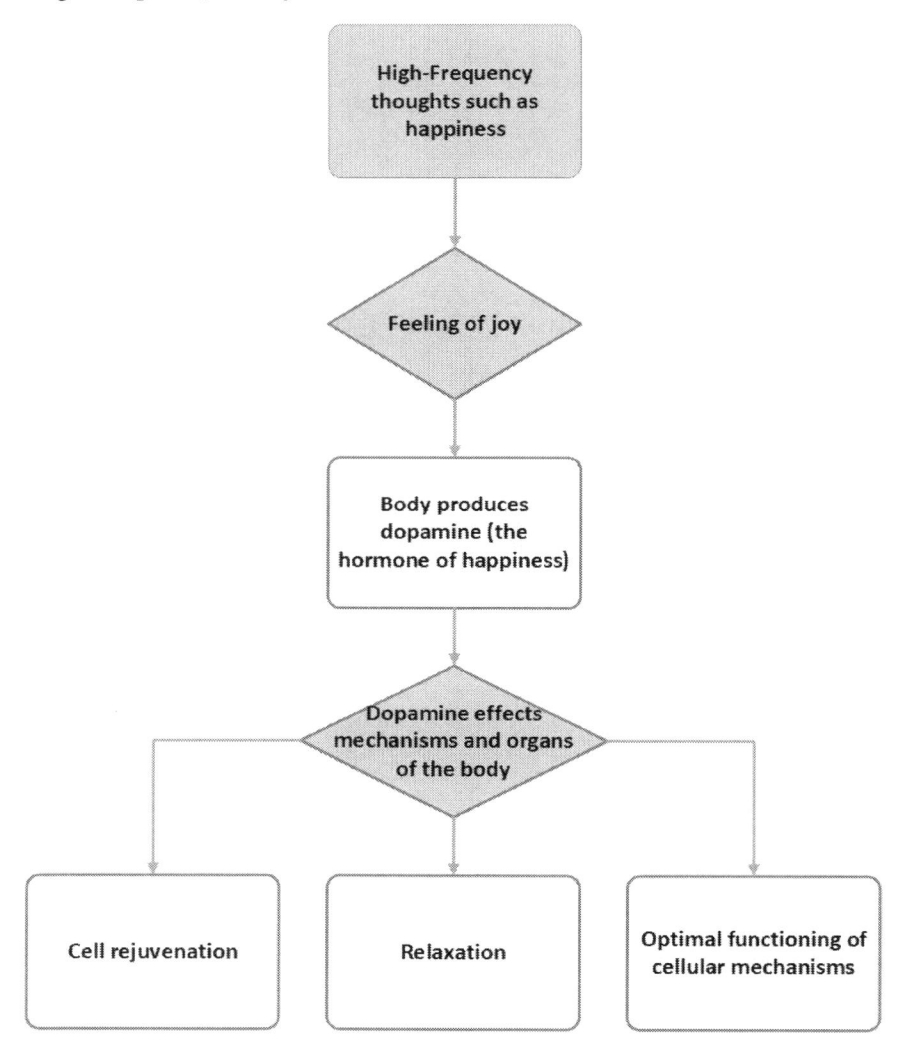

High-frequency thoughts:

```
┌─────────────────────┐
│   High-Frequency    │
│   thoughts such as  │
│     happiness       │
└──────────┬──────────┘
           ▼
        ◇ Feeling of joy ◇
           │
           ▼
┌─────────────────────┐
│   Body produces     │
│   dopamine (the     │
│ hormone of happiness)│
└──────────┬──────────┘
           ▼
    ◇ Dopamine effects ◇
    mechanisms and organs
        of the body
   ┌───────┼───────┐
   ▼       ▼       ▼
┌──────┐┌──────┐┌──────────────┐
│ Cell ││Relax-││   Optimal    │
│rejuv-││ation ││functioning of│
│enation│      ││  cellular    │
│      ││      ││ mechanisms   │
└──────┘└──────┘└──────────────┘
```

THE TRUTH ABOUT LIFE

In truth, behind every illness there are negative thought patterns and unhealed resentments that need to be released. Behind every illness there are negative feelings such as unforgiveness, anger, guilt, or rage that need to be released. Behind every illness there is an emotional pain that needs to be released. Behind every illness there is a divine message, a spiritual message that needs to be addressed and released. Pains, imbalances in the body, and illnesses are simply signals that some parts of life need positive changes.

The truth is that diseases and health conditions carry messages or lessons that need to be uncovered and learned. Paying attention to the part of the body where there are imbalances and where the diseases have manifested will give clues regarding the message that your soul is trying to convey to you. Diseases are like metaphoric messages or signals coming from the soul to bring into our attention specific issues, emotionality, or thought patterns so that we can address and correct them.

The following study illustrates the fact that diseases are definitely associated with emotionality and subtler issues beyond the physical: In a scientific study conducted by Timothy W. Smith, Ph.D. of the University of Utah and published in the *Journal of Psychosomatic Medicine* in 2011, the relationship between some hostile marital behaviors such as anger during disagreements, low marital quality, controlling and dominant behaviors during marital interactions were measured and scored in 154 couples with no prior history of cardiovascular diseases. The study analyzed the relationship between the hostile marital behaviors listed above, the risk of coronary artery calcification, and the risk of coronary artery diseases (CAD) in these couples. The results of the study showed that hostile behaviors such as anger during disagreements, low marital quality, controlling and dominant behaviors during marital interactions are directly correlated to a higher rate of coronary artery calcification and to a higher risk of developing coronary artery diseases (CAD).

This study clearly shows that hostility in marriages is associated with an increased risk of cardiovascular diseases. Marriages as well as other love relationships involve the "heart", not the physical heart but the emotional and the spiritual hearts. When people use the term "He broke her heart," they are not referring to the physical heart, but everyone knows what a "broken heart" means. It is with no surprise at all then that Dr. Timothy W. Smith's study showed that displaying hostilities and harsh behaviors in marriages is associated with hardening and calcification of coronary arteries and to an increased risk of cardiovascular diseases in couples involved in these marriages.

A few years ago, one of our friends, who I would call here James to preserve his anonymity and who was a physician, had great difficulties in his personal life. In fact, during that time, he had marital issues with his wife and was also so strained at work that he felt like he couldn't breathe. He spent countless hours in a stressful job as a physician, and every time he got back home, he couldn't find peace there either. He felt like he was trapped, suffocating in life, had no respite, and couldn't breathe wherever he turned, whether at his work or at home. How big was my surprise when one day we heard that our friend James had a pulmonary embolism, couldn't physically breathe, and was rushed to hospital. James's story is just another example among countless cases that shows that there is more to the causes of diseases than what we currently know. In truth, the part of the body where a disease has manifested gives valuable cues about the real causes of the health issues involved.

Examples of Health Issues and their True Meanings

Below are few examples of health issues and the possible signals and messages that they point to:

Health conditions	True causal issues	Changes or actions to undertake in order to heal
Liver problems	Anger	Releasing anger, letting go of judging others or situations, working on yourself to attain self-control, poise, and acceptance
		Forgiving people and life situations
		Letting go of the feeling that people are bad and unjust or that the world is unfair
		Please check the chapter about "ego" to gain more understanding about anger
Back pain	Carrying too many charges, responsibilities, and burdens in life	Lightening your schedule, delegating some responsibilities to others, taking time to rest
	Doing too much on your own	Setting down the heavy burdens such as financial burdens or an overly busy schedule, heavily loaded household chores, and anything that may be weighing on you or be too much to carry on your own
	This health condition is a metaphor for carrying too many burdens or charges on your shoulder or on your back	
Wrist pain	Clinging to unhealthy things and situations in life	Finding out the unhealthy things, people, or habits that you are clinging to in life and simply deciding to let them go
	These unhealthy things may include:	
	-Toxic relationships that you know are not good for you but still you cling to them	
	-Careers that are not appropriate for you, but you don't want to let them go	
	-Destructive habits and	

	behaviors that you are clinging to or anything else that you are clinging to and do not want to let go	
Eye problems	Related to the third eye It is associated with not being able to see the spiritual truth Being unable to see the truth about life or situations Being blindsided by one's own opinion and not being able to see others' perspectives Being spiritually blind (forgetting who you are and literally walking in the dark in life and being lost) Losing touch with the truth or the reality of life Seeing other people as separated from you (below, inferior, or superior to you) instead of knowing and seeing the Oneness in all people	Seeing everyone as divine precious beings equally valuable as yourself Being willing and open to see the truth about situations, yourself, and your life Seeing and understanding the big picture and the truth about life Being willing to see love, worth, and value in everyone Seeing clearly spiritually and being aware of the spiritual beings, your guardian angels, and your guides that are around you
Uterus-related issues	Related to the sacral chakra imbalances Being in an unhealthy or abusive relationship with a partner and not being willing to let go of the unhealthy relationship Having past sexual traumas or sexual abuses Associated with emotional issues with one's parents, especially one's mother Holding negative feelings and judgments toward your mother, for instance	Letting go and releasing any negative love and sexual relationship that you are in Forgiving past lovers, partners, and spouses Finding psychologists to help with past emotional issues and healing from any sexual trauma or abuses Forgiving and loving your parents: father and mother Letting go of childhood traumas or any negative feelings related to your mother, father, ex-lovers, or current lovers

Being overweight	Being overweight is a sign of protection. It is related to building a barrier around yourself in an attempt to protect yourself against people, harsh energy, or anything else Can be understood as unconsciously building up a wall of energetic barrier around yourself, which then manifests in the physical body as stacking up fat and adipose tissue around yourself as a protective barrier	Finding a way to feel safe in life Letting go of people, situations, and things that are toxic to you and do not make you feel safe Asking God and your angels to protect and help you to feel safe in this world Using discernment and intuition to select whom you allow in your life Knowing that life is safe and trustworthy
Heart conditions	Related to the heart chakra and the spiritual heart Can be due to several factors such as: having a fright, being abused in the past, or someone that you love deeply was removed from your life circle—and that created pain in your heart Not loving yourself and others	Dealing with the underlying causes accordingly and releasing the causal emotions Open your heart to love again after relationship breakups Forgiving all abuses and betrayals Learning to love yourself and others Releasing pain from your spiritual heart chambers through the processes of forgiveness, self-love, and trust

As mentioned above, diseases and pains are signs that some parts of your life need positive changes. Diseases arise from negative thought patterns and emotionality that are then expressed in the physical body as imbalances, and they carry messages or signals that point to the underlying causal factors. Paying attention to the location where the disease has manifested in the body will give you valuable clues as to what needs to change in order to heal. For instance, heart diseases point to the spiritual heart issues, while hand-related health issues point to holding on to unhealthy things or jobs. Therefore, the best approach will be to pause at the first sign of a slight discomfort in the body, ask for guidance, and listen to the message that the body and your soul are conveying to you. You may ask the following questions: "What is this health issue here to teach me?" "What is my body trying to show me?" "What is my soul is trying to tell me?" Then, once the underlying emotions are revealed, they will be released and will dissolve into the ether. Then the body will return to health.

The ultimate healing remedy, though, is learning to love through the process of forgiveness and living in joy. In truth, healing, when truly understood, means: "casting a strong decision in the direction of your Higher Self's priorities." What this means is that when you are deviating from your true self's priorities or your soul's priorities, you will experience pain and discomfort (dis-ease) as a way of pointing you and redirecting you to the right path. It is like a feedback system where negative feelings and diseases serve as negative feedback showing you that you are deviating from love, from the truth, and from your life's purpose. On the other hand, joy, happiness, peace, and a healthy body serve as positive feedback showing you that you are on the right path. Therefore, removing all negativities and anything that does not serve you from your life is the path to health.

In fact, negativities lower the frequency of the body and lead to disease, but many people are not even aware of that truth. These negativities come in various flavors and textures such as unforgiveness, anger, jealousy, guilt, shame, gossiping, sarcasm, judgment of yourself and others, toxic relationships, dramas, stress-filled lifestyles, addiction, drugs abuse, junk foods, criticisms of others or yourself, seeing separation by seeing others as inferior to you or not valuable, not being true to yourself and to others, junk feeding of the mind with unnecessary dramas, etc. All these negativities, with repetition, decrease the frequency of the cells, and the final stop is the manifestation of disease in the body. On the other hand, anything that is aligned with love, kindness, compassion, forgiveness, being on your life's path, or soaked in love will lead to pure, vibrant health.

The body is made to thrive in love, and when exposed to negative frequencies of our thoughts, emotions, the environment, and the garbage of the mind, the vibration of the body's cells is disturbed, the cells' frequencies are lowered, and this eventually leads to diseases.

I would like to point out here that the information provided in this section is solely for the purpose of gaining greater awareness and understanding regarding the truth about health and healing and that there is no need to feel guilty or blame yourself if you are going through some health challenges. Know that you are always doing the best you can with the information that you have available at each moment. Therefore, there is nothing to feel guilty for. Furthermore, guilt is very detrimental to health.

The Biggest Unknown Truth about Diseases

Is There Anything Else Regarding the Cause of Diseases That You Would Like to Share?

Yes. The biggest unknown and untold truth about diseases is that some health problems are related to past life events or issues, especially those that are resistant to medical treatments. For instance, if someone has water phobia and is afraid of swimming, it is really probable that this person may have drowned in water and died in one of their past lifetimes and is carrying the traumas of that tragedy in their subconscious memory. That person will not necessarily know that dying by drowning in water in a previous lifetime is the real cause of their phobia and fear of water. This is true for many other phobias as well. In fact, we carry memories of past lifetimes with us, and these memories can interplay and surface and cause physical or emotional illnesses in our current lifetimes. In truth, many health issues are related to past life experiences.

Let's suppose that someone was a cruel king in an ancient past lifetime, who cut the hands of his servants who did not perform their tasks well; because he was seen as an invincible king and worshiped by his subordinates, he did whatever pleased him during that lifetime. After his death and life review, this person, who was so caught up in the illusion of life and falsely believed that he was an invincible king, may choose to experience what it feels like to have his hands cut and live without both hands in his current lifetime. Since we are no longer in a period of our evolution on Earth where cruel kings cut the hands and feet of their subordinates, that person may develop a disease such as cancer in the arms or may be involved in a car accident or suffer from some other diseases that will leave him with both of his arms cut so that he can experience the lessons that his soul has chosen to learn. This can be called paying back a past life karma, but, in fact, it was a decision of the soul itself to experience what it is like to lose both arms as a lesson to learn from past lifetime choices. In truth, there are many health problems related to past-life issues that people are not aware of. In many cases, these health issues related to past lifetimes are hard to treat or simply resistant to medical treatments. One option to heal from these past-life-related health issues can be to do regressions in order to release these emotions and traumas. Once the causes, the traumas, and the emotions are revealed and released, the body will follow to health.

Below are a few other suggestions that can assist with the process of healing:

Meditations	Regression
Visualization	Change in diet
Prayers	Forgiveness
Affirmations	Living in love and joy
Energy medicine	

Light of Truth

As a scientist and pharmacist, I was shocked when I first discovered the truth about health and illness. I asked myself: *"Can this be really possible? Can all the health issues that we experience in life really come from holding on to negative thought patterns, toxic feelings, unhealed emotions, and disconnection from love and past life issues?"* Alas, it is true, and I am a living example of that truth. In fact, I suffered so much with various health issues for decades, but after I started applying these principles and healed myself, I can't help but tell the truth. Indeed, I was able to heal from decades-long diseases, which physicians had even told me were incurable and that my only option was to manage them with medications. However, I have used the principles described in this chapter to heal myself from all the health issues that I used to have. Not only did I start the process of forgiveness and clearing all negativities in my life, I also started a series of affirmations, visualizations, meditations, and prayers. I blended all these techniques together and applied them in my life, and I found healing for my body and peace in my mind for the first time in my life.

Later, I discovered to my amazement that I was not the only one who has unexpectedly discovered the truth about health and healing. There are millions of people who are currently applying these same principles to heal themselves from several health conditions ranging from the slightest diseases such as the common cold to the more serious such as cancers. For instance, Dr. Joe Dispenza, in his book *You Are the Placebo*, shared numerous recorded cases and showed with scientific evidence that it is our thoughts, beliefs, and emotions that determine our life experiences, health, and happiness. Today, he conducts workshops across the world and uses scientific technologies to measure biological changes in the body induced by thoughts.

Another example is the case of Louise Hay, the founder of Hay House. Indeed, after healing herself from cancer by using visualization techniques, Louise wrote several books on the topic of power of thoughts, affirmations, visualizations, and forgiveness in the healing process.

Moreover, Dr. Bruce Lipton, former medical school professor and research scientist, and author of the bestselling book *Biology of Beliefs*, laid the foundation in his book that supports that it is the environment of the cells, not the genes, that controls the destiny of cells and that thoughts and beliefs do initiate and induce biochemical changes in the body. He pointed out with the support of scientific discoveries that the body can change as we retrain our thinking. Currently, physicians and other healthcare professionals are interested in his scientific discovery and assist his seminars and teachings to gain a better understanding.

Another undeniable example that speaks for itself is the case of Julie Renee Doering. In fact, Julie, who is today the #1 on brain rejuvenation, survived the Atomic Bomb testing radiation poisoning in the Nevada desert as a child. Julie Renee has had 17 surgeries, multiple cancers, and at a time was in a wheelchair and was told that she would never walk without a cane. However, by using visualizations and meditations, she succeeded in rejuvenating her cells and healing herself. Today, she is helping countless people around the world to heal by teaching the technique of cellular neo-genesis and brain rejuvenation.

These are only few examples. In fact, there are countless scientific studies and recorded cases in the fields of neuroscience, psychosomatic, quantum physic, cellular biology, epigenetic, etc., that support the evidence that thoughts, beliefs, intentions, and our emotions do impact and affect our health. Do not just take my word for it; try it yourself and you will be amazed by the self-healing capability of the body and the power of thoughts, beliefs, and visualization in the healing process. Every health condition is healable. It is truly possible to heal from all diseases. I am sharing this truth today because I know what it is like to suffer from health issues. I have been there. Everybody is capable of healing and living totally healthy. I mean everybody. Health is your divine birthright. Do not give up.

Children and Health Issues

How Do You Explain the Cases of Innocent Babies Who Have Terrible Health Conditions and Sometimes Even Die From These Diseases? What Negative Thought Patterns Do They Have or What Messages Are They to Learn?

It can be hard at first to believe the real causes of diseases as explained in this chapter because of how we are trained in our society. The mind will automatically come up with many objections, questions, and several scenarios in an attempt to dismiss the truth. The first question that the mind will automatically come up with are: *"What about babies or little children who are sick? What have they done wrong or what thought patterns do they have or what messages are they to learn?"* I asked these same questions myself when I first discovered the truth about health and diseases. However, before answering this question, I would like to reiterate here that no one has done anything wrong. Being sick does not mean that you have done something wrong or incorrectly. There is no right or wrong doing here; diseases are simply ways that point to changes that need to be done in life and can be seen as learning tools. Diseases can even be perceived as your soul speaking to you to get your attention in order to help you on your earthly journey. That's all.

Coming back to the question of health issues faced by young children, let's examine the case of a baby who has a heart failure condition as an illustrative example. When there are imbalances in the body such as a heart condition in the young baby, for example, this is coming from the soul of the baby that has been called forward to retrieve and retract from the physical world. Let us just say that the baby transitions (dies) due to heart failure. This is the soul of the baby calling itself back into pure consciousness. That is not a physical problem; that is a spiritual calling. However, many humans, particularly the parents of the baby, when faced with this, would be in grief, a sorrow so deep that can't even be named, and they would not see it this way. They would just say, "My baby has died of a heart failure." However, this is the ego mind thinking because the spiritual truth is quite different from our human perceptions.

You may ask, "Why did the baby decide to die (transition into pure Consciousness) at such a young age and through the means of a serious disease such as heart failure?" The answer to this question may be that the soul Consciousness (of the baby) chose to come into a lifetime that will be brief and has served its purpose. It is now ready to go on another adventure of its evolution. For the most part, this will happen when the parents or the family of the baby need certain life lessons. The stay is brief, and the baby retracts. The soul is eternal. This must be understood by everyone. A soul is eternal, pure Essence, and has no body, no sex, and never dies. It is pure Light. It is inconceivable for us, humans, to understand that a soul is pure Consciousness and does not need a body in order to exist. When the baby decides to leave, it is so that the parents or caretakers will receive soul lessons from the baby's departure (death).

You may ask, "What kinds of lessons are needed to be learned by the parents of the baby through such a hard experience as the death of their child?" Well, the lessons can be a variety of things such as forgiveness of self, forgiveness of life and God, how to transform suffering into light, or how to love yourself and fall back in love with life again even after you lose your child. What often happens is that there is great blame placed on the human heart by ego when a baby passes. The humans, the parents, blame themselves. They grieve and lose touch with the understanding of life. These lessons will be learned through the baby's passing. Had the baby stayed and grown up into adulthood, these lessons would not have been learned this way. At times, suffering is the teacher.

What about Children Who Have Diseases, yet Don't Die from the Disease but Suffer All Their Lives? What Is the Purpose of That?

When you sit closely to a child who is in a condition of embodiment of a disease, there is a certain quality vibration that you will feel; you will also see the Light Consciousness through their eyes. There is something special about these children. But, of course, the purpose of this is to fulfill their own desires for growth through an evolution. They have chosen it. Other human beings, particularly the parents of these children, are the beneficiaries of the teachings. It is also a great sacrifice, one that should be revered. These children have made great sacrifices, and it takes great courage to make these life path choices for their incarnations. These children have selected this, and those who are the beneficiaries of the selection are blessed. The spiritual lessons of the mother and father of such a child are marked with a high light, a high vibrational consistency higher than most because the suffering is great.

There will come a time where there will be less suffering leading toward the teachings and more joyful lessons that are learned out of joy because the axes of our planet Earth are now changing, and harmony is presiding. There will be, in the future, fewer diseases and more joy emanating from our life lessons on Earth. Don't be fooled by appearances; when you sit closely to a child afflicted with diseases, you feel the Light emanating. It is a teaching and a blessing for all.

How Can You Call This a Blessing When the Parents of These Children Are Suffering?

It is normal to have this question, but I am talking about deep spiritual concepts and spiritual perceptive here. Therefore, we are talking about something that is beyond the human mind and beyond human perceptions. The parents are suffering because they are connected to the human mind; there is identification with the human mind. What do I mean by that? The suffering comes from identifying with the human body, forgetting that you are eternal, and the whole earthly experience is like a drop of water in an ocean compared to the evolution the soul. A parent might say, "My child is ill and will remain ill for the rest of his life. My whole life is ruined." However, what this parent is calling "my whole life" is like a drop of water in the ocean compared to all the lifetimes and the existence of their soul, and the soul of this parent knows that very well; so does the soul of the child. That is why in the Spirit world, with this knowingness and awareness in mind, both the parents and the child did not hesitate to make this decision and planned a lifetime where disease and suffering would be the teachers. However, above the human mind experience, we observe and we see Light. It is for each parent to discover for themselves the

Light that surpasses the human mind experience and these lessons will be learned either in the physical experience with their child or when they have passed into the nonphysical. Either way, the teaching is inherently good. It's human to say, "How can this can be a blessing when I am seeing my child suffer?" That is a trick of the ego mind. I am not discrediting the weight and the heaviness of the suffering; I am saying that there is something beyond the human mind. The suffering is created through the thinking, but, beyond that, there is an opening to experience Light.

I can't even imagine how difficult this can be, and I am very compassionate. I also know that there is something that lies beyond all of these. I invite parents to look and see what that is. Just be willing to look. It is possible for the parent of a child suffering from a disease to feel the influx of great Love and to reside in that Love beyond the suffering. It is possible. These are not just words; it is an experience, a spiritual experience like walking on hot coals. This can be seen as if these parents decided to walk on hot coals during their incarnations and yet find a way to move beyond the suffering and to feel love and peace during that difficult experience. *Why have they decided that and what lessons are they to learn in these?* It is for each parent to look and find the lessons and the blessings. *Why did Jesus incarnate fully knowing that His body would be nailed on a cross at the end? Why are we all here in this difficult planet to learn to grow, and to learn to love and forgive?* Because we all know that, at the end, only blessings and lessons will come from our experiences and that we are eternal, and at the end, all is well.

For more information on how to heal and to learn more about health and disease, please refer to my other book, *Health and Disease Prevention*.

The Truth about Life
CHAPTER 4

THOUGHTS, BELIEFS, CREATION OF REALITY

"Truth is powerful and it prevails." –Sojourner Truth

Thought, Beliefs, and How to Release Negative Thought Patterns

Is It True That We Create Our Lives and All the Experiences in Life with Our Thoughts?

Yes. This is one of the greatest truths about life that many are not aware of. There is a truth that if you think, it will be so. These thoughts that we think are like droplets falling into the water. First, there is a little puddle that forms, and then a great lake. Oh yes, something will develop out of your thoughts. There are still people who do not believe or understand the truth that we are building our lives as we go through our thought energy. This truth cannot be emphasized enough because most people are still riddled with negative thinking.

Thoughts are like seeds that you plant in the subconscious mind. We create with our thoughts, words, and feelings. Thoughts are not just thoughts; even if you cannot see them, they are energies; they are vibration, and they carry specific frequencies, and scientists are now able to measure the frequency of thoughts with the advanced technologies that we currently have on Earth. These thought energies go out of the mind, associate and magnetize with other thought-like forms, and attract to your experience people who have the same types of thoughts or are going through situations similar to your thoughts. Every thought that you think creates a thought-like form that has the potential to manifest into the physical world as objects, experiences, or situations. Your thoughts are like magnets that attract to you things, people, and situations that you repetitively think of. Therefore, becoming conscious and selecting your thoughts is crucial.

There is no neutral thought. A thought is either positive or negative. Since every thought has the potential to manifest and appear in the physical world, we should be watchful of our thoughts and select thoughts that will help create a positive and beautiful life.

For instance, every object that surrounds us was once a thought from someone's mind. Tables, clothes, houses, cars, art, highways, airplanes, plates, as well as everything that we use on a daily basis were first ideas and thoughts that came from people's minds. Indeed, for each of these examples, someone first thought about the idea or concept, designed it in their mind, and then and it became a reality that was created into the world as a material object. Every thought has a creative potential and power, whether it is positive or negative. Keep in mind that even the idea or the thought of creating atomic or nuclear bombs was conceived in someone's mind. This insane thought of creating atomic bombs was then acted upon; then atomic bombs were created, which led to the catastrophic disaster of World War II. Therefore, we should be very careful in selecting our thoughts and not fantasize about negative or crazy thoughts unless we want to see them manifest in the physical world.

Can You Explain a Little More about the process of How We Create with Our Thoughts?

Here is how it works: Repetitive thoughts become a belief; the belief then creates and attracts experiences based on the original thought blueprint. For instance, let's suppose someone thinks they are not likeable. If this person thinks this original thought long enough, emotions and feelings that match this thought will be created, and such a person will start to believe that people don't like them. This will cause them to mistrust people since they believe that people do not like them. When in the presence of other people, such a person will be more likely to display behaviors that match their belief such as stepping back from people, or isolating themselves, or closing their heart to others. People will sense or feel that, and since no one really likes people who are withdrawn and closed in on themselves, people will be more likely to dislike such a person, which will reinforce the original belief that people don't like them. This thought and belief will become a self-sabotaging and a self-fulfilling prophecy, and the vicious circle will continue. That is how thoughts become beliefs and how beliefs will draw into your life things or situations that you believe in.

Here is another example: When you repetitively think that you are not smart because your parents, teachers, or others have told you so or for whatever reason, you will start believing this idea, and it will become your experience and then manifest in your life. This is how it works: When you believe that you are not smart, you will be more likely to be filled with doubts about yourself and lack self-confidence. When faced with a test at school, for example, you may panic, and your brain will go into fear mode. Consequently, the sympathetic nervous system, which is the part of the brain involved in fight-or-flight reactions, will be activated, while the other parts of the brain involved in intellectual activities will become frozen during your test because you are in doubt and in panic mode. Therefore, you will be more likely to perform poorly on your test and may get a bad grade. This bad grade will reinforce the original belief that you have about yourself of not being smart. This belief will set up the momentum for your future tests, in which you will probably perform poorly as well. Furthermore, when you are studying, you will be more likely to get easily discouraged and not motivated, and you will be more likely to dislike studying because of your previous experiences of getting poor grades. After all, who would like to spend time doing something that they know is not paying off or will turn out in a failure as before?

Consequently, you will be more likely to give up in front of the slightest difficulty, and you may say to yourself, "*This is too hard for me. I am not good at this.*" Since you have already brought home some bad grades from school, your parents and teachers may say something about your poor performance at school, which will only reinforce your belief. If you do not check your beliefs and find a way to change them, you may grow up into adulthood with these negative and limited beliefs and may develop other related beliefs similar to your original belief of not being smart such as not being competent, or being average, or not being good enough, and so on. The belief of being average or not being good enough will reinforce the concept that you have about yourself. Consequently, you will be more likely to perform at an average level in whatever you do in your workplace, family life, parental role, etc.

How We Create Our Lives With Our Thoughts:

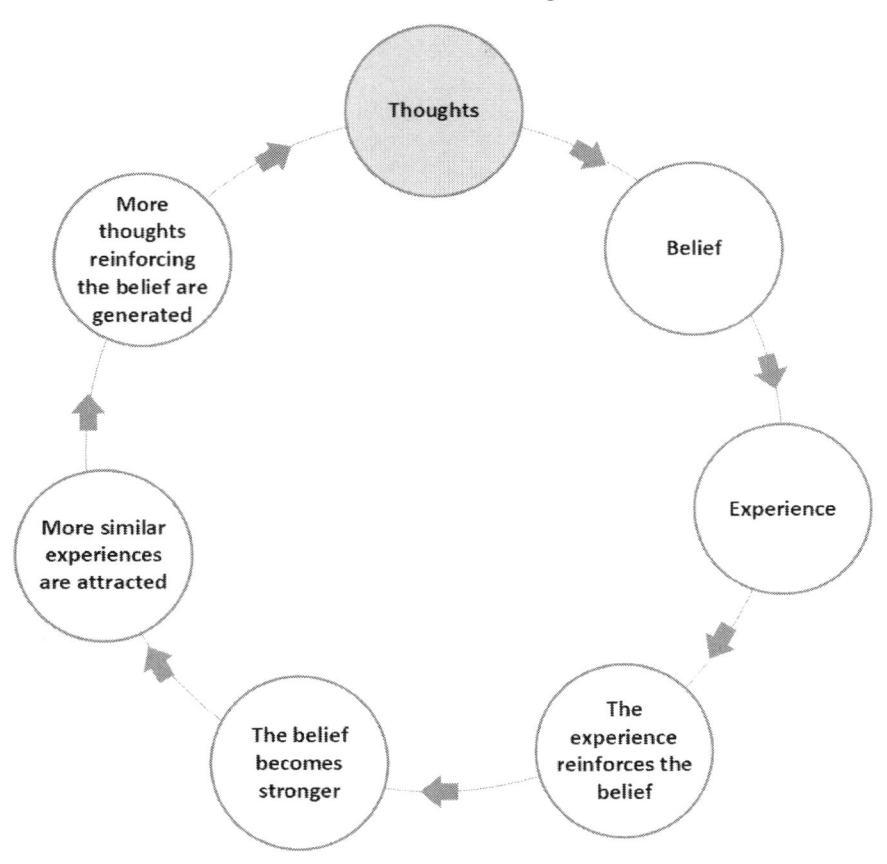

- Thoughts
- Belief
- Experience
- The experience reinforces the belief
- The belief becomes stronger
- More similar experiences are attracted
- More thoughts reinforcing the belief are generated

Can You Give Us Some Concrete Examples of the Power of Thoughts and How They Create Experiences in Life?

Here Are Some Examples of How the Circle of Thought-Belief-Experiences Works:

<u>Example 1:</u>

A woman thinks often that she is unworthy and not loveable because of childhood experiences and because she was rejected by her mother and abandoned by her father when she was young. This repetitive thought becomes her belief. She grows up and meets a guy in a nightclub, and she is very attracted to him. She has a few red flags that this guy may not be such a great guy, but she was blindsided by the love euphoria of a first-time relationship, and since she has a belief that she is unworthy and not loveable, her expectation is not so high (she was not expecting Brad Pitt or Bill Gates to propose to her anyway), so she is just happy to find someone who loves her and wants to be with her. The honeymoon period wears off quickly, and her new partner starts to treat her poorly and disrespectfully. She is shocked and does not understand what is going on. However, it was her belief that attracted her to a love partner who treats her poorly and who does not respect her. This poor treatment and abuse from her partner further reinforces the belief that she has about herself of being unworthy and unlovable. Because her belief is reinforced, she now thinks that she must be truly unworthy and unlovable. She now has more unworthiness and degrading thoughts about herself. Consequently, she is more likely to stay in the unhealthy relationship or attract similar relationships unless she changes her thought patterns and her negative beliefs.

Example 2:

Let us say that someone thinks that they are superior to others or their race is 'the superior race', which, by the way, is nothing but false thinking and a crazy idea because we are all One and simply take on different body-vehicles that will best suit our life purposes. But this is the thought of this hypothetical person, and this thought becomes his belief—they believe that they are superior to others.

Let's take the case of a hypothetical person; let's call him James in this example. James is a white man who thinks that he is superior to others and his race is 'the superior race' because of his education or for whatever reasons. This repetitive thought becomes his belief. Now he sees the world through the lens of this false belief that filters his experiences, magnifying and attracting to him experiences that reinforce that belief. James regularly watches TV and sees in the evening news the dramas, the burglaries, gun shootings in the ghettos, and all this foolishness committed by an infinitely small fraction of people who are lost in life. However, what he watches and hears in the news reinforces his belief that the other race must be truly inferior. Because his belief is reinforced, he now thinks that he must be truly superior to the other race since he is not a burglar, has never committed any infraction, and has had a secure job as a gatekeeper in a local corporation for more than a decade. This belief attracts to him other like-minded friends and people that he meets in bars, at his workplace, and other places. James and his new friends often have discussions about what is happening in the news and all the troubles that the other race is creating in the society. One day, this belief attracts to James a situation in a restaurant with a black waiter who was having a hard day and who inadvertently misplaced his order. This minor mistake, which may have gone unnoticed by other people, really angers and enrages James, especially coming from a black person, again reinforcing his belief that the other race must be really stupid or inferior. This incident with the black waiter creates only more thoughts about his superiority to the other race, which reinforces even more his original belief and attracts even more similar experiences and situations to him. The vicious circle will continue unless he changes his thought patterns and his limited and self-inflated beliefs about his superiority to the other race.

Ways to Change Negative Thought Patterns

How Can We Quickly Change and Assimilate Positive Thinking?

Even though changing thoughts and beliefs is a process, the first step and the best way to start is by quickly switching your intention to something positive every time you find yourself thinking negative thoughts. One way of doing this is to have the imagery of something positive, something of a high vibration, an image that you can burn into your mind whenever a negative thought arises. I will suggest finding images that will be like reference pointers so that every time you find yourself thinking negative thoughts or having thoughts about the past, you can quickly switch to these images or things that will make you smile or feel more at peace and happy because these imageries or references are of higher vibration and thus are more positive frequency thoughts.

Here are a few examples of how you can use your imageries to stop the train of negative thinking in the mind:

Example 1:

When negative thoughts come into your mind, you can quickly switch your thoughts and imagine yourself being like a dolphin swimming in a beautiful clear blue ocean, free of all worries, all life's struggles, and concerns; you are just swimming free as you are light, relieved of all burdens, and peaceful. This beautiful blue ocean also has a powerful healing energy that is healing you and appeasing your mind as you are swimming.

Example 2:

Every time you see the rise of negative thoughts in your mind, you can imagine yourself on a beautiful beach, sitting on a peaceful bench with Jesus sitting by your side. Jesus says to you, "Everything is going to be okay," and an indescribable peace that surpasses all understanding fills you and both of you are just sitting there on that peaceful beach, looking at the sunset. You are at peace, relaxed, and just breathing and watching the sunset with Jesus beside you.

You can do this imagery exercise with any other religious figure that resonates more with you or to whom you are close to such as the Buddha, Moses, etc.

Example 3:

Every time you find yourself thinking negative or hurtful thoughts, you may also think about Archangel Michael. Imagine that Archangel Michael is watching over your life right now and working behind the scenes to help you with everything.

Archangel Michael is a powerful, 100% unconditional loving angel, with a reassuring, protecting, and calming presence. He completely understands our human conditions and has great compassion for everything we endure. He is here to help humanity and will help everyone who calls upon him, no matter their religious backgrounds. You don't even have to believe in angels in order to garner his help; he is here for everybody, will never judge you, and unconditionally accepts and loves everyone, including you.

Every time you have negative thoughts, you can quickly switch your thoughts to Archangel Michael and imagine that He knows your name, your situation, and is working behind the scenes to support you.

Besides using the imageries, another helpful technique is to take a breath and center yourself and your focus on the present moment by turning your attention to the things or people that surround you whenever you find your mind drifting into the past or negativities. It is a way of turning your attention to the here and now and what is happening in the present moment instead of being lost in the thoughts.

Using references and centering on the present moment are simple and quick techniques to break free from the thoughts. However, in truth, these two techniques will only give temporary relief from the thoughts; for a lasting change of the mind and thought patterns, deeper inner work will be required. I will say that these two techniques are simple and quick ways to break out of the spin of negative thoughts and stop their momentum by not giving your attention to them. However, to truly heal the thoughts, these techniques must be accompanied with deeper inner work and mindful exercises such as meditation, forgiveness, affirmations, self-love, and understanding the truth about the thoughts in the head and disengaging from them. Meditation is crucial for getting out of negative thinking. Getting out of negative thinking is a real process, but it is worth it. It is a win-win process because you'll get to feel good in the process, and every minute devoted to learning to think positively is worthwhile.

Below are some inner work practices:

- Meditations
- Learning coping mechanisms for dealing with negative thoughts
- Affirmations
- Identifying your belief systems and writing affirmations to counter them to create new, positive beliefs
- Writing down the thoughts that continually run in your mind; this will reveal to you how negative and false these thoughts are
- Doing forgiveness work
- Seeking human help with psychologists or counselors if there have been traumas
- Praying and asking for divine assistance
- Reading positive and uplifting spiritual books

Light of Truth

Everyone can change their thoughts and everyone is capable of doing this. There is no situation too messed up, and nobody is too far down the pit of despair that they can't start changing their thoughts and lives. For instance, a good friend of mine, Tonia, had a very challenging and dark past. In fact, she grew up in a dysfunctional family, was sexually abused for several years when she was young, was a drug addict, was involved in prostitution, later lost one of her children, and was even homeless for many years. However, she succeeded in changing her life; she got out of the road, started working on herself, spent a lot of time in meditation and self-healing, and today she is a successful international businesswoman with several employees running a multimillion-dollar business. As one of her dear friends, we discuss often and I know for a fact that most of her success has come from learning to change her thoughts and letting go of the past. She has found a way to heal herself and only live in the present moment and not in her thoughts or in the past. Today, she helps others to get out of addiction and teaches how to let go of negative thinking and the past. Today, she is happy and living an abundant life.

Now That We Know That It Is Our Thoughts and Beliefs That Create Our Experiences, How Can We Change Our Beliefs?

Let us first clarify what a belief is. In fact, a belief is nothing but repetitive thoughts that we think so many times or that our society or parents tell us so many times that we finally accept them as true. Beliefs are also concepts and ideas that we have collected through our own experiences in life that have become so anchored in the subconscious mind that we accept them as truths. If you imagine thoughts to be like droplets of water, repetitive thoughts become like a lake, and a belief is when this lake of thoughts becomes so anchored in the subconscious mind that it becomes crystallized and forms ice. That's why a belief is so powerful and also challenging to change.

The trick is that many times we are not even aware that we have these beliefs in our subconscious mind and that they are running our lives. For instance, many people believe that in order to succeed in life, they must get a degree then a job and slowly climb the echelons of their companies. This is just a belief, which, by the way, is a little limiting because many of the richest people on this planet such as Bill Gates and Oprah Winfrey did not take that preplanned flat road. However, the majority of people believe this limited idea, and it has become their experience, and many work in corporate business all their lives with no respite. As you think it, you will believe it. As you believe, it shall be so for your life.

Now that we know what beliefs are, how can we change them? I will say that the first step is to do a little bit of investigation in order to know what beliefs may be running in the subconscious and shaping your life that you may not even be aware of. But how can you change something if you don't even know what it is or if you are not even aware of its existence? How then can you change your beliefs if you are not even aware that you have beliefs or when you don't even know what they are? Thus, doing a little enquiry and checking your beliefs is helpful in order to change them.

Any thought that you believe, and it does not matter whether the thought is helpful or absurd, the simple fact that you are focusing on it and holding that belief will stir the wheel of manifestation, and it will eventually manifest in your life. If you have challenges in some areas of your life, and if you do a little enquiry, you will find out to your greatest surprise that underneath you have some beliefs that are creating or attracting these challenging situations to you. Therefore, it is useful to take some time and check your beliefs in several areas of life. Doing so will reveal to you hidden beliefs that may be in the background and running your life unconsciously so that you can work on them and change them.

Below are some areas in life where you may consider checking your beliefs. They include some questions that you can ask yourself in order to access the beliefs that you may be holding.

1. Relationships (this area includes friendship, relationship with coworkers, love relationships, family relationships, and so on)

Relationships	Limited belief	Positive belief
	Do you believe that women are complicated, gossipers, emotional, speak too much, and often melodramatic?	Or do you believe that women are calm, speak little, are kind to each other, and impassive?
	Do you believe that men are mostly untrustworthy?	Or do you believe that men are trustworthy and faithful?
	Do you believe that the majority of marriages and love relationships have trouble and that it is just the way it is (just complicated)?	Or do you believe that it is possible to have a drama-free, loving, honest, and peaceful marriage and loving relationship?
	Do you believe that most people are fake and double-minded?	Or do you believe that most people are honest and trustworthy?
	Do you believe that girls or women are catty and difficult to get along with?	Or do you believe that women are peaceful beings who are easy to understand and to get along with?
	Do you believe workplaces are places of competition filled with	Or do you believe that workplaces are peaceful places

THE TRUTH ABOUT LIFE

	people who complain, gossip, and stab each other in the back?	filled with loving, supportive, and trustworthy people?
	Do you believe that every family has issues and that there will always be relatives that you cannot stand?	Or do you believe that it is possible to get along and have peace with everyone in your family?
	Do you believe that it is rare to find good and trustworthy friends in this world and that having best friends is limited to childhood and high school times?	Or do you believe that trustworthy people and friends exist in this world and that you can meet and make great friends at any age in your life, no matter how old you are?
	Do you love and accept yourself unconditionally?	Or do you believe that you are unworthy, have messed up so many times in life, and that you are not good enough?
	Are you able to forgive others no matter what they have done?	Or do you believe that some mistakes are unforgivable and some people are simply jerks who do not deserve to be forgiven?

2. Career

Career	Limited belief	Positive belief
	Do you believe that you have to work hard to earn money?	Or do you think that you can earn your life with ease and grace and can access abundance effortlessly, smoothly, and without struggle?
	Do you believe that you are worthy and deserving of a great career?	Or do you believe that your fate is sealed and that great, joyful, and fulfilling jobs can't just seem to find your door?
	Do you believe that work in general is a chore that we all have to go through in order to earn our living?	Or do you believe that a job can be enjoyable and fun while making great money—in other words "Employment is Enjoyment"?
	Do you go to work just to pay your bills and get by in life?	Or do you work each day in joy while getting the best out of it and making great money at the same time?
	Do you work for the sake of earning your life?	Or do you know the natural talents and gifts that God has given you and work in a job that is in alignment with your talents and gifts?
	Do you go through life with no definite purpose?	Or do you know your life purpose? (It may or may not be related to your work.)

3. Money and abundance

Money and abundance	Limited belief	Positive belief
	Do you believe that money is the source of all evil?	Or do you believe that money is a good thing?
	Are you jealous of those who are rich?	Or are you happy for those who have money?
	Do you think of yourself as being poor or lacking money?	Or do you believe that you are abundant?
	Do you constantly complain, wanting more money?	Or do you feel grateful for what you already have?

4. General concepts and beliefs about life

General beliefs about life	Limited belief	Positive belief
	Do you believe that life is unfair?	Or do you believe that life is fair?
	Do you believe that life is hard?	Or do you believe that life is easy and benevolent?
	Do you believe that life is against you?	Or do you believe that life is for you?
	Do you believe that your circumstances are playing against you and that you aren't just "that lucky"?	Or do you believe that you have the ability to accomplish anything in life and that success is within reach?

Once you discover or uncover your limited beliefs in each of these areas, the goal will be to replace them with new positive beliefs by using affirmations. Then all you have to do is write down affirmations that reflect the new beliefs that you want to introduce and repeat the affirmations daily. With time, and as you practice the affirmations, your beliefs will begin to change, and your life will start to change as well; you will draw new, positive experiences and people into your life. For instance, if one of your beliefs is "Life is hard," you can write an affirmation such as "Life is easy and benevolent, and I float easily and effortlessly in life," or you may write your own.

These are only a few areas just to give some ideas on how to check your beliefs. You can do the same exercise in others areas of life to discover and change the beliefs that you have in such areas. Other areas in which you can check your beliefs may be: heath, spirituality, God, and so on.

Affirmations as a Powerful Tool

You Mentioned Using Affirmations to Help Change Our Thoughts and Belief Systems. Can You Expand More on That? How Do Affirmations Work?

When the thoughts become very negative, they become very compelling, and it is not easy to let go of them or just drop them instantly because we become used to them and the thoughts have gained momentum. Therefore, to break free from the vicious circle of the spinning of negative thoughts in the mind, the first step will be to change the negative thoughts into positive ones.

One way of breaking free from negative thought patterns is through repetition via the use of autosuggestion to introduce new beliefs in the mind. Autosuggestion is another name for affirmations. For instance, for people who believe that they are not smart, the first step is learning to autosuggest to themselves that they are smart. Keep in mind that this first negative belief of "not being smart" has probably been suggested or introduced into their minds by their parents or teachers or they have created these beliefs by repeating to themselves over the years that they are not smart. How many times have they probably told themselves words like, "I did something stupid again," or "I always mess things up," or "I am not good at this or that," or simply "I acted like a dummy," or "I am not just that smart," etc.? All this negative self-talk gets stocked up in the subconscious mind and convinces them that they are not smart and will get reinforced every time they make little mistakes or do something incorrectly. Since they are adults now, and there is no parent around them to tell them that they are smart, the only way of reversing this belief is taking charge of their own minds and to start autosuggesting to themselves that, in fact, they are smart, well able, and can do anything.

THE TRUTH ABOUT LIFE

These people must write new affirmations and repeat them several times a day for as long as needed until their subconscious mind starts believing in these affirmations. Keep in mind that, at the beginning, their subconscious may not believe these affirmations because they are too foreign or too far from their old habitual negative ways of thinking, but just like when you are learning to bike, it may seem difficult and impossible at first to find your equilibrium; with repetitions, the mind will begin to believe the new affirmations. Thus, they will have autosuggested or introduced new beliefs into their subconscious minds by using affirmations.

When you affirm or listen to something repetitively, with time, your subconscious mind will end up accepting it as true. This is an unfailing law that is always at work and always true. For instance, many researches in psychology have shown that what parents tell their children repeatedly in their childhood can tremendously affect their states of mind and shape their adult life experiences. This is because the subconscious mind is like a sponge that absorbs ideas, thoughts, and beliefs presented to it. These beliefs will be carried throughout life unless the person proactively takes charge of their mind, makes some inquiries, and changes the beliefs to new, positive ones. Indeed, the mind is very malleable and will soak up whatever information, criticism, or praise is suggested to it. This information, whether it is good or bad, true or false, will go into your mind, and whatever program enters the mind will then manifest itself through the adult years.

In the same way, you can modulate and shift your beliefs by intentionally and proactively using positive affirmations and repeating them constantly. When you affirm something, your thoughts are focused on them. While the ideal is not to think so much and live mostly in the present moment, the truth is that when the mind becomes very negative and filled with criticisms, unforgiveness, judgmental thoughts, hurt, anger, and all kinds of negativities, the first step is turning these thoughts into positive ones via the power of affirmations. To be honest, negative thoughts are like sticky glue, and they are not easily eliminated when they have their grip on the mind, which is why they must first be changed into positive thoughts before even considering letting go of all thoughts and living mostly in the present moment.

The goal of affirmations is to help change old belief systems that are useless and detrimental into new, positive belief systems. Affirmations are like planting new seeds in the garden of your subconscious mind, which will grow. These seeds will become your new truths because as you say them repeatedly, your subconscious will assimilate them as truth.

Unfortunately, many people are not even aware that they are creating their lives through their thoughts and beliefs and never take the time to enquire why the same patterns, the same types of people, or the same type of situations keep coming into their lives over and over. Consequently, they never take responsibilities for their thoughts and often feel victimized. They simply go through life, are tossed here and there by the winds of life, and then they blame their circumstances, their parents, their bosses, their coworkers, the government, God, life, or whoever of being responsible of their misfortunes and hardships in life. The truth is that everyone is creating their life with their thought energies and beliefs. This is why everyone must take responsibility for their thoughts and work to change their beliefs and the energy that they are sending forth through their thoughts, words, and actions.

This is not to blame anyone because when you realize that you are creating your own life with your thoughts and beliefs, it is possible to fall into self-accusation, feel guilty, or blame yourself for your life. I would say, instead, be grateful that you now know the truth, and you can now work on yourself to change your thoughts and beliefs so that you can create new experiences. Indeed, 97% of the world's population are not even aware of this truth or have never heard that "thoughts and beliefs create and shape life." Therefore, consider yourself fortunate if you are reading these words; this means that you have been searching and are ready to change your thoughts and to create a beautiful new life for yourself.

The principle of autosuggestion is the same principle used by marketing experts, and it is the basis of the commercials that we hear or watch on radios and television. They play the commercials over and over and, with repetition, we end up believing the messages behind the advertisements. I remember a song that I heard on the radio a few years ago. At first, I thought that the song was not that great. Every time I turned on the radio, I heard the same song day after day. One day, I found myself singing the song and even thought that the song was not that bad after all because I had heard it so many times and become accustomed to it. Affirmations work in the same way: with repetition, your subconscious mind will start to believe them, even if they seem untrue to you at the beginning. The goal of affirmations is to use the same principle that marketing experts use to change your thought patterns and to introduce new beliefs that can benefit and change your life in a positive way. This means that it is never too late because you can always change your thoughts. When you change your thoughts, your life will change as well.

How Do You Concretely Change Your Thoughts and Beliefs with Affirmations?

The best way of doing this is to write a series of affirmations that are tailored to the thoughts and beliefs that you desire to change and start repeating them daily. You may write them down, read them aloud, or record them and listen to them daily. In fact, I personally do all of these: I have affirmations that I wrote on my phone, which I can read anytime. I've also recorded some of my affirmations that I listen to daily.

The affirmations have to be written in the first person and in the present tense such as, "I am," "I have," "I do," etc., in order to make it believable to the subconscious mind. Saying, "I want," will keep you in the vibration of wanting, which is a state of lack because if you want something, it means that you don't have this thing, you are lacking it. For instance, instead of saying, "I want money," it is better to write an affirmation such as, "I am grateful that I am financially free," or, "Money flows to me all the time, and I am abundant in all ways." If you desire to write affirmations related to relationships, for instance, instead of saying, "I want love," or, "I want great relationships," it is better to say, "I am loved. I am now surrounded by loving and supporting people. All my relationships are joyful and loving."

It usually takes at least 40 days for the brain to start rewiring and change. I will suggest that you continue doing the affirmations and make it a daily practice even when the 40-day threshold has passed and even if your mind starts to change or you start to see positive results in your life so that you can keep the momentum going.

Light of Truth

I have been using affirmations for several years now, and I can say from my personal experience that they are amazing. Growing up, no one taught me about the power of thoughts, and I lived a life in which I kept attracting negative situations and hardships one after another, and I did not even know that it was my own thoughts that were filled with fears, resentments, anger, and negativities that were sabotaging my life and attracting these situations to me. One day, I had a mental breakdown and realized that I could no longer live that way and started to work on myself using different tools including affirmations, deep inner forgiveness work, meditations, and regression to learn about myself. I had many limited and negative beliefs that I was not even aware of until I started to do some inquiry and dig further. In fact, I knew nothing about the power of thoughts and was being tossed from situation to situation and often wondered: *"Why did this happen to me?* or *Why am I always attracting these types of situations and people into my life?"* I didn't even realize that it all had to do with my thoughts, beliefs, and the vibration that I was holding.

As I started to do my personal inner work and change my thoughts, my life began to change. I started to draw new loving people into my life; even strangers that I met in grocery stores were nicer to me. Today, I rarely encounter angry drivers on the highways. I was even reconnected to family members that I had not spoken to for years, and my relationships became more harmonious. I was amazed by the results, and I can say from my own experiences that if you really change your thoughts, your experiences will change, and your life will change too; it has to change because if you change yourself, your external circumstances cannot stay the same, they will change. It is a law. It is possible to live in peace and harmony in this world, and I know people who are living this upgraded type of life; they have discovered the truth about thoughts and are applying that truth in their lives; they are happy, peaceful, and joyful. It does take some work, but the result is worth every single minute and every effort devoted to it. If you do the work, it works.

The Truth about Life
CHAPTER 5

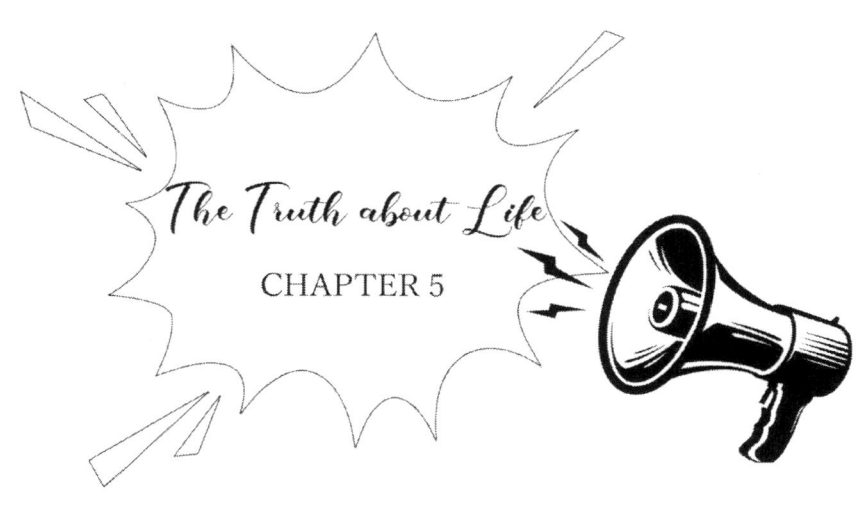

THE EGO

"You will know the truth and the truth will set you free."

—John 8:32

What Is the Ego?

The ego is the voice in our minds that speaks all the time, which we mistakenly take as our own voice. The ego is the thoughts in the mind. We all know what it is like to be doing something and yet not be focused on what we are doing because our minds are somewhere else, distracted by the thoughts in our head. This is exactly what the ego is: the constant chattering in the head. Sometimes, the ego can be so loud that we may be doing something but are lost and sometimes not knowing what we are doing or making mistakes because of the distracting thoughts in our mind. These thoughts, these voices in the mind are what I am referring to as "the ego". Some people refer to it as the ego, while others refer to it as the egoic mind, the voice in the mind, the thought screen, the monkey mind, and so on. I will use any of these terms interchangeably to refer to the voices in the head. Every human has these voices in the head, and the great majority of humans believe that these voices in their heads are their own voices. However, one of the greatest truths about life is that we are not the thoughts in our heads. These voices in the head come from activities of the brain, like an automatic computerized software that automatically goes on and starts doing what it is programmed to do—talking, chattering, commenting, criticizing, comparing, complaining, bringing out past stories, stirring up past arguments by feeding them with more thoughts, imagining worse fearful possibilities about the future and worrying about them, blaming, bringing up guilt and shame, and the list goes on and on.

One of the biggest challenges of the human experience is these voices in the head and coming to know that they are not our own voices. How is this sophisticated computer in the mind created and why do we incarnate with it? I will say that you can see the human experience as coming to a school to learn how to choose love over fear and to learn to love. To make this school more complex, you come with a computer in your mind, and you never knew it's a computer, but you believe it to be you, your own thoughts. This sophisticated computer in your mind will often say things like, "Look what others did to you. You need revenge. You need to lose weight. You are poor. Nothing is going right in your life. Get back at those who have hurt you..." However, your role in this school is to come to a point where you realize and say, "No, I am going to choose love. I will forgive. My future will be good. I am not going to do anything that is unloving because this is not who I am," constantly choosing love on a daily, minute-by-minute basis, even though the ego in your mind tells you to do otherwise.

THE TRUTH ABOUT LIFE

This schoolroom of Earth can be seen as though you are placed in an environment like in an experiment where the goal is to learn to love and to choose love, but there are 'people' all around you who are inciting you do otherwise and to go astray, away from love. However, these 'people' are not real people but voices in your own head that are constantly bombarding you with stories, scenarios, and ideas that are unloving, pushing you into doing unloving things or criticizing you and others. Once you give in and do an unloving thing, the same voices then flood you with guilt, shaming you for doing so, and will say, "Look what you have just done again. You always do that…" To make this school even more real, you have come to believe that you are the voices in the head since the voices are coming from your own mind, and it seems impossible for you to guess that they are not the real you but just an automatic software producing thoughts in the brain, which is programmed to talk, and its talks are mostly contrary to love and are untrue. To make the school even more complex and interesting, you do not even know that you are in a school, neither do you know how you got here in the first place and who you truly are. Therefore, most people live their entire life in a sort of mental prison, lost in their thoughts, getting bullied by these thoughts, and falsely believing that they are the talking voice in the head. You may ask, "Why do we have the ego in our minds or criticize its purpose?" but doing so would be like enrolling yourself in a nursing curriculum and then wondering why your clients are all sick people and why you are not treating only healthy people. In other words, the voices in the head, the ego as I call it, are an integral part of this Earth curriculum that you have elected to enroll yourself in. So, everyone on Earth has this egoic mind that talks, and, in fact, the goal is to eventually realize at some point in your evolution through your many incarnations that you are not "this voice in the head" and to choose love even if it is telling you otherwise. The ego comes in various tones and may take different roles, but it is the same ego playing these various roles.

Various Roles Often Taken by the Ego

Below are some roles that the ego or the voice in the head often plays.

The parent:

This is when the ego starts telling you things such as, "Did you do the dishes?" "Did you close the door when you left home this morning?" "Have you cleaned the bathroom?" "Did you brush your teeth this morning?" Indeed, the ego will sometimes act if it were your parent and give you unsolicited parental advice. Even though it may seem like the ego is trying to help you, notice that everything that comes from the ego will make you feel bad.

The buddy:

Here, the ego will come up with a story to distract you; it talks to you like a buddy, telling you funny things or reminding you of things that you watched on TV that made you laugh. When listening to the 'buddy' you may be performing your task and you may be smiling and laughing on your own. This may seem like an inoffensive thing when the ego takes on this role, but they are distractions that keep you involved in the mental realm of the ego instead of being present and enjoying the present moment.

The coach:

Here, the ego starts to act like a coach, as though it was helping you and saying things such as, "You need to lose some weight," "You are eating too much," "You need to act more elegantly when you are around people." However, you will notice that these sentences coming from the ego will make you feel bad and are not helpful at all. In fact, nothing good ever comes from the ego because the ego always has an agenda behind it and only pretends to be helping you.

The commander or the dictator:

This is when the ego starts to give you orders such as, "Hurry up; you are going to be late for work or school," "Work faster; you are too slow; you are getting behind in your daily task," "You need to finish your to-do list."

The blamer:

Here, the ego comes up with all the things that you have done wrong in the past or could have done better and fills you with guilt. In this case, the ego will come up with sentences such as, "You always mess things up. What is wrong with you?" "You are not smart enough. How could you have done that?" "What were you thinking?" "This is so dumb!"

The enlightened one:

Here, the ego takes on the role of superiority and tells you that you are better or superior to others. It may say things such as, "You are better than others," "Others are stupid," "You are enlightened," "You are prettier than others," "You are richer than others," "You are more educated than others," "You are more refined," or "You have more properties and fine manners than others."

Besides taking on these various roles, when the ego presents itself in the second person with words such as, "You need to do this or that. You always do this or that," and so on, notice that oftentimes the ego talks in the first person using words such as, "I," "me," "mine," and this is what makes it even more believable and it fools us by making us think that we are the ones who are thinking, without realizing that it is just the voice of the ego.

Light of Truth

When I first realized this, it was as though something lighted up my mind. I was stunned upon realizing that the voice in the head is not mine but a computer that talks. I came to a point where during deep meditation, the voice stopped, even if it was just temporary. This was when I realized, *Woaaooh! These thoughts are only but empty voices that can be stopped through meditation practices, and they are not me.* In breathing meditation, sometimes, I can see the thoughts flashing in my mind. I would say that the only way to fully understand that you are not that voice or the ego is to start meditating. It is then and only then that you can fully understand the grip and the falsity of these thoughts that we all have in our minds. It is something that every human has to discover on their own, not something that you can read about and then the ego will disappear or dissipate. In fact, it may take several years of practice and meditations to come to a point where the ego is relinquished in the background and you fully live in the present moment most of the time. Actually, it may take 10 years of practice and often it takes many lifetimes and incarnations before waking up from the illusion and recognizing the ego for what it truly is. However, every second dedicated to quieting the mind, meditating, is worth the trial, as it will calm you down and help you to relax. In fact, I know people who live that way, mostly in the present moment, and their egos are relinquished in the background of the mind. Actually, I have some friends who have attained that level of spiritual evolution. It is possible for all. As for me, I still have these voices in my mind, but I am not controlled by them anymore and do not follow what they say as I can see them for what they are.

The truth is that when you follow the voice or the commandments of the ego, you will be unhappy and often stressed out because the ego will give you assignments and tasks to do; unless you realize that and say, "NO," to it, it will push you to do a lot of things and then blame you later for doing them.

The ego is also the one that urges people to talk a lot. When others are talking, the ego will push you to think about the next best reply to say. The ego will convince you that what you have to say is so interesting that it can't wait and that you should say it immediately. One of the ego's principal signatures is to make you to think that the things that you have to say or do are so urgent that they can't wait. Those who live in their egos often talk a lot too, not realizing that it is the voice in their heads that is urging them to speak too much. That is why, sometimes, when people come back from silent meditation retreats, they learn to recognize and detach from the voice of the ego that urges them to talk. Sometimes, the only way to realize this is to do a silent meditation or decide to spend one or two days without talking and without speaking a word. Once you make that decision of not talking, you will see how many times the ego will urge you to talk and to say something. The more you learn to say, "No," to the urges of the ego to speak, the better you will become at keeping quiet and being present and listening to others.

Why Do We Have Ego?

The main purpose of the ego is that it was first designed to protect us when faced with danger. This was useful at the time when humans were living in tribes and there were dangers, wild animals and frequents attacks from other tribes. At that time, the ego was useful to alert that danger was ahead or to help to recognize the enemies. Therefore, that is why the ego operates mainly out of protective instinct and out of fear. The ego is the basic, primitive part of the brain that operates from fear, protection, and survival. However, in these modern times, humans do not live in forests anymore. Therefore, nowadays, the ego mainly creates unnecessary fears, stress, anxiety, and worries; it sees dangers where there are none.

Another purpose of the ego is that it was designed to see separation. In fact, we incarnated with these computer-like minds so that we can experience life as "individual beings separated from each other." Because the ego was originally designed to see separation, it now only sees division, fear, separation, and lack everywhere. Today, the ego will not say, "The person in front of you is the enemy or from the other tribe," rather, it will say, "The person in front of you is different from you. He is from another race. He is black. He is fat. He is ugly. He is poor," and so on.

You can see the egoic mind as an outdated computer that is no longer needed because we can live without the voice, the commandments, the labels, the judgments of the ego in our mind. In fact, many are living that way currently on Earth. Indeed, our true selves do not speak with thoughts but with intuition, knowingness, and feelings, gentle urges to take action, and so on. The biggest challenge that we face when we incarnate in this third-dimensional world is that we believe we are the ones thinking, while in truth it is simply the ego in the mind that is talking; thinking is just happening. This is maybe the biggest illusion of this third-dimensional plane and the biggest challenge as well. It takes some time and practice of meditation before realizing that you are not that voice in the head and before overcoming this illusion.

Other Characteristics of the Ego

1. Thinking about the past or the future

The ego is the voice in the head that is always thinking about the past or the future because when you are in the present moment and in the here and now, the ego will fade away because the mind cannot do two things at the same time; you are either in the present moment or lost in the thought screen of the ego. In truth, the only moment that exists is the present moment.

2. Classifying people into categories

The ego is also always afraid and sees the world and the people around as, "Me against them." It does not see unity, and it knows nothing about Oneness and love. In the ego world, everyone and everything is classified as good and bad, tall and small, rich and poor, me and them, beautiful and ugly, and so on. The ego also says that people are superior or inferior to others or are better or worse than others and so on. The ego or the voice in our minds creates so much pain and suffering in this world because we think and believe the voice and act upon it. However, the point is not to blame the ego because, like a small child, the ego does not know better. Compassion and acceptance are what are required here, plus knowing that the voice in the head is not you but simply a computer designed to talk and classify things and people.

3. Liking and disliking

The ego likes to use sentences such as, "I like this," or, "I don't like that," "I like this person and I hate that person." The ego or the voice in the head likes using words such as "I," "mine," "me," and it always want to be the center of attention of the world. Every time you use the word "I," know that your ego is back in the front line.

4. Making assumptions

The ego is likes to make assumptions, pretending that it knows everything, but, in spiritual truth, no one knows what life will bring in the future.

5. Telling half-truths

The egoic mind and the thoughts in the head are not true: they are simply lies or half-truths. What makes the ego's lies very believable is that most of them are half-truths. For instance, let's say that someone lost his car key one day and was very upset because he was concerned about being late for work. Let's suppose that it had happened to him a few times in the past where he lost his key or other items. The ego or the voice in his head may tell him, "You always lose your key," or, "You always lose your stuff. You are not organized," which makes him feel bad. This is a half-truth because, of course, he does not always lose his items, and it happened to him only few times in the past. More than 99% of the time, when he wakes and prepares to go to work, he does not lose his key. Just because he lost his stuff a few times one cannot say that he always does that and label him as an unorganized person. Half-truths are not truths but simply lies. However, they are really believable. That is how we often get caught in the lies of the ego; we believe them and start to feel bad about ourselves.

Identification with the Ego

Most of us are identified with the ego at least to some degree. In fact, our whole planet is under the epidemic of the ego, which means that many go through life thinking, thinking, and thinking. Many do not even know what the ego is and are not even aware that the thoughts in their heads are not theirs but the ego's. Deep identification with the ego is the source of suffering and pain, and truly it is the cause of all the problems, wars, and all the issues that we see on Earth. There are various degrees of identification with the ego, and the more identified you are with the ego, the more you will suffer and the less happy you will be in life. However, there is no blame here. Be where you are, do your best to become more aware of your ego; do the best you can to get out of the ego.

Below are some signs of identification with the ego.

Anger (which comes from judging)

Since the ego only sees separation, and sees others as different, separated, and apart from it, one of its favorite activities is judging: judging others, situations, and everything. The ego only sees things as white and black, good and bad, and it knows nothing about the concept of spiritual evolution and that everyone is doing the best with their level of spiritual maturity. Therefore, the ego likes to judge others as being bad, wrong, not doing well, not doing good enough, etc. because it believes that others can be better than what or who they are or can behave better. Thus, it judges them. You may ask why if it is these people themselves who are judging others and not their egos, but here is how it works: Let's say that someone is rude to you. The ego may say, "How can he do that? How undiplomatic and careless he is. This is not right …" and if you agree with these angry thoughts, before you know it, you will be immersed in the ego's world, and because you agree with the thoughts and believe that you are the one thinking, the ego will fuel it with more angry thoughts, and you will become even angrier. That's how identification works: you will start to believe that you are the one producing all these angry and unloving thoughts. On the other hand, if the ego comes up with angry statements, and you say, "No, I am not going there. I choose love. He is doing his best, and I am not going to criticize him," the thoughts will cease. As soon as you believe even a little bit in the ego's statements and agree with them, the ego will continue bringing these angry thoughts into your mind.

What many people do not know is that anger comes from judging others and situations and that anger always follows judgment. What do I mean by judging? Judging is when the ego says, "This is wrong!" "How can he or she do that?" "Why do they do what they do?" "This is not right!" "This is unfair!" "This is white and this is black!" "This shouldn't be this way!" "This shouldn't be happening right now!" "This is a no-no," and so on. Every time the ego judges, if you believe it and agree with it, you will feel angry. In fact, anger is a sign that indicates that you are believing something that is not true spiritually and that you are being caught in the ego's lies. Therefore, those who are deeply identified with the ego are often angry. However, those who are less identified with the ego or do not live in their egos are more tolerant and more accepting of others. The truth is that everyone is doing the best that they can with their level of spiritual understanding and maturity. Therefore, there is no wrong or right action; people are just advancing at their pace and with the tools that they have.

Judging others is very similar to asking, "Why does a nine-month-old baby crawl and why is he or she not walking?" and being angry about that. However, the truth is that a nine-month-old baby cannot do any better than crawl, and it is the best the baby can do with his or her age and level of growth. However, since the ego does not know anything about spiritual truth and cannot and does not want to understand the Oneness and sees only separation, it judges others, and if you believe and agree with it, you will be angry. The more you believe and go along with the ego's judgments, the angrier you will get about people and situations. Furthermore, the ego does not only judge others but will also judge you. Indeed, the ego will judge you and tell you that you are not doing well enough and that you are bad, or that you are this way or that way, etc., and if you believe that you are these thoughts or you are "this thing" who is judging yourself and others, you will feel bad about yourself and you will suffer. You may even develop low self-esteem if you are deeply identified with the ego and fall into its lies.

Stress:

One of the biggest weapons of the ego is creating unnecessary stress. Those who are stressed out are at the mercy of the "dictator ego". Indeed, the dictator ego likes to give assignments; it is what it does. It will tell you that you need to finish your entire to-do list or that you need to take care of or find solutions to something that is going to happen tomorrow or in the future. Therefore, those who are deeply identified with the ego and who do not know that it is the ego that is giving them these mental tasks and assignments will be running around nonstop, trying to finish all tasks on their endless to-do lists and will not even have a second during the day to pause, breathe, or rest. In fact, the ego's assignments and bossy orders will never end unless you recognize the trick of the ego and say, "No, this can wait. I can do this tomorrow or another day. I don't have to think about or take care of this right now."

On the other hand, those who are less identified with the ego or at least who are not listening to the "dictator ego" are those who are relaxed, calm, and peaceful in any situation. In fact, it is not that the ego does not give them orders or tell them to rush, but it is just that this second group of people simply somehow come to know that these thoughts in their head are not true, and when the "dictator ego" tells them, for instance, that they need to do something or rush, the simply internally reply to the ego, "I don't care." Therefore, they are more relaxed and peaceful in life because they don't listen to the lies of the ego or act upon the orders and assignments that the ego gives them.

In fact, "I don't care," is a sentence that I have used a lot and continue to use until today; it helped me tremendously to get out of the dictatorship of the ego. Indeed, I used to be stressed out about several things, but after I discovered the truth about the ego and its tricks, every time the ego tells me, "Hey, you need to hurry up or you are going to be late for work or a show or an appointment, etc.," I simply reply to the ego, "I don't care." Surprisingly enough, I have become more and more calm and peaceful and the dictator ego's demands and suggestions have started to fade away with time.

Below are other signs and symptoms of identification with the ego:

Always wanting to be right	Guilt
Believing that you have one and only one life	Resentment
	Anger
Fearing death	Believing that your life will end with your death
Believing that you are the labels or you are the "roles" that you are playing in life	Being stressed
	Fear
Worry	Discontentment

Unhappiness
Diseases
Impatience
Talking a lot

Jealousy
Being competitive
Believing that you are your body
Etc.

Getting out of the Ego

How Can I Release Myself from the Prison of the Ego?

First, you must understand that the thought in the head is not the true you. It is the shadow, the chatter, the labeling in the head. In fact, it is an activity of the mind that can be measured through scientific equipment. It is like a barking dog. When we realize that this is not the true self, it is then far easier to observe this. This untrue activity in the mind is a habit; that's all. Watching and observing as you are self-blaming, self-criticizing, and self-harming yourself with the thoughts in your head will help you to realize that it is not you but the voice of the ego. Watching and observing is more than enough; this will dissolve the patterns. Observation starts with the feeling of one's own presence, which is cultivated through self-love. (Please, if you need to, go back and check the section regarding self-love in the relationships chapter.) It is interesting that when you love the self, it will immediately help you to understand that you are not the self-blaming/judgmental judge in your head, which is nothing but the ego mind. It is a form of activation. By the way, when you realize and wake up, the light bulb goes on, and you realize that you are not that voice that blames and harms and criticizes you; that is a barking dog that has been trained since childhood to bark, and you can remove yourself from it and watch the barking dog. It is the ego mind that is barking.

The activation or illumination will occur in each human and some point of the evolution when they will finally recognize that there is a false sense of self and a true self. One of these is real (the soul), the other is illusion (the ego). Every human being must pass this test and, throughout the duration of their lifetimes, every human will be given an opportunity to do so. To break free from the mental prison is to realize that "that voice" is not you, it is a barking dog, and to be able to observe it. Then and only then it will fade away on its own. Some humans have big dogs that bark very loud, and for a long time, a lifetime, they live with this barking noise in their heads. Others' egos are smaller and easier to quiet. The truth about life cannot be more real than that. The fundamental truth about life must be to know oneself truly and to clearly identify the false self—the ego self—and to release it.

Are There Ways to Get out of the Ego?

There are ways out of the ego, and the ego can be at least managed and even put in the background. This happens when you live in the present moment most of the time and no longer listen to the voice and the half-truth of the ego. There are some ways and practices that can help to get out of the ego's world and its frantic demands or at least manage the ego to an extent so that it does not take over the driver's seat and run your life.

The starting point is to come to the realization that you are not that talkative commentating voice in the mind. This awareness itself is a huge step and helps you to detach so that you can become the observer of the thoughts, observing the ego do what it does and yet noticing that it is not the real you.

Meditation

The first technique and perhaps the most important tool that helps to get out of the ego is meditation. Remember that the ego is the voice in the head that constantly talks. Therefore, it is during meditation that we can practice quieting the mind and see the thoughts for what they truly are: half-truths and lies. It is during meditation that we can train ourselves to quiet the mind and live in the present moment. It does take some practice and taking time to meditate every day, but it is all worth it. I want to point out here that when you start meditating and working on quieting the mind, the ego will try to become your helper on this journey and will tell you, "It is not going fast enough. Look, you are thinking again. Have you done your meditation today? You need to do your meditation." The ego will stress you out if you let it, unless you are able to detect its schemes and realize that your ego is being "reborn" into something else—"the spiritual teacher"—and ignore its commentaries.

Practicing the art of silence

Consciously practicing to be quiet and talking when it is only necessary is another way of overcoming the ego. Practicing the art of listening to others and being present with others is a helpful means of getting out of the ego. Remember that the ego is what always pushes us to talk and it also wants us to believe that we have something exceptionally interesting to say and share with others. It urges us to talk while others are talking and to think about the next thing to say. Although there is no urgency in sharing any news, the ego will make you believe that people have to hear what you have to say, and they must hear it now. Therefore, practicing the art of silence and saying, "NO," when the ego or the voice in the head asks you to talk and share something is really powerful, and the more you do it, the better you will become at keeping your mouth shut, listening to others, and talking less.

THE TRUTH ABOUT LIFE

Listening to sounds around you in silence without thinking

Listening to the sounds around you in silence without thinking about the sounds but just listening as they emanate is another technique that is really helpful in noticing thoughts and quieting the mind. I often set some time off to listen to the sounds in my home, just sitting and listening without thinking. It is amazing how much sound you can hear when you sit silently and listen to the sound in your environment. It will make you realize that the voice in the mind is really noisy and blocks you from hearing other things. I can't agree more when some people call the ego "the noise in the head." The ego does make the mind foggy.

Giving to others

Another thing that can help one get out of the ego is giving to others. Remember that the ego is always afraid and always wants security, and it does not like to share. The ego will make you believe that if you give, you will not have enough left for you. However, this is not so; when you give, you will automatically receive because the universe will find a way to give back to you. Remember that the ego operates from survival and protection and is concerned for its survival in this world. Since the ego sees others as separated from it, it cares more about its own welfare than giving and helping others. Therefore, giving to others is the way out of the ego. When I say giving to others, I mean giving money, your time, your talents, doing something good and nice for free to help others.

Practicing loving kindness

Another powerful tool to get out of the ego is to practice loving kindness toward everyone. You can do this by loving and accepting people just as they are, no matter how they are behaving. You have to decide that you want to be kind and that you are going to get there no matter what. In fact, it takes a strong commitment and decision to say, "From now on, I will do my best not to judge anyone in my thoughts, words, and actions." The truth is that you will fail many times, but you have to forgive yourself and get back on track. It is a practice, and I can attest that it does work. When I started this work, I literally wrote down the type of person I wanted to become and made it one of the goals of my life, just as one of my goals was to finish my education. I worked on this day and night and even prayed for help and assistance. I can testify that it does work, but first you have to make the decision and be willing to change and stick to it with great determination. Otherwise, the ego will take over your mind and run the show.

Seeing the divine in others

Since another toxic playful time of the ego is criticizing others and everything, you have to make a firm decision to see others with love instead of seeing only their physical appearance, which is a merely an illusion. Therefore, focusing on seeing the divine in others and seeing others with love is another useful practice to adopt. One thing that I often do, which helps me a lot to see the divine in others, is calling everyone that I meet "a child of God." I often say, "Here is another child of God," when I meet a new person. Doing this will change your perspective and the way you see people; it will help you not to focus on negative things such as noticing how fat or ugly they are, how they look, and so on, which the ego loves to do.

Light of Truth

The truth is that no one will get out of the ego by their own strength but by Grace; thus, no one can boast about it. It does take practice and patience to come to a point where you completely live in the present moment most of the time. It can take up to 10 years and often it takes many incarnations. Stories of people who wake up one day and their ego just disappears and the voices in their mind stop are rarely true stories; or if they are true, it is probably because these people have already worked on the process of getting out of the ego in their previous incarnations. However, the simple fact that you are reading these words means that you have come to a point of spiritual evolution where you are ready to put the ego in the background. Be patient and gentle with yourself throughout this process because anything worthwhile takes time. The truth is that everyone will get out of the ego at some point, either during this lifetime or in future lifetimes, and when that happens, you will live completely in love and light and will no longer need to come back here on Earth unless you choose to come back to teach and guide others. You will become "enlightened". This was the case of some beings such as Jesus, Saint Germain, Moses, the Buddha, and others who came back to teach, guide, and serve. However, meanwhile, every effort done in the direction of quieting your mind and getting out of the ego is worthwhile. Moreover, you will get to feel good in the process as you will become more loving and peaceful.

To have a better understanding of what the ego is, please refer to the book *The Power of Now: A Guide to Spiritual Enlightenment* by Eckhart Tolle. This is a book that I have read and I believe that it explains very well what the ego is.

The Truth about Life
CHAPTER 6

REINCARNATION, KARMAS, DHARMA, AND CHOICES BEFORE BIRTH

"No legacy is so rich as honesty." –William Shakespeare

Reincarnation

Do We Reincarnate?

Yes, we do.

Can We Reincarnate as Animals or Trees and so On?

Life is all about evolution and progression. Although all is possible, as human beings living in a third-dimension plane we often come back as humans several times until we progress and evolve into other higher dimensions such as the fourth or fifth dimension. It is all about progression and evolution, not regression. Humans are more evolved than animals and trees, so they will evolve in a higher form or dimension, not the opposite.

Do Karmas Exist, and How Can You Explain Them?

Unfortunately, the word "karma" is often misunderstood, and there are many misconceptions around it. Indeed, the word has got a bad reputation in our world and is often association with punishment, the Divine spanking you for bad behavior, paying back for mistakes, and so on. However, karma is simply a lesson that you have signed up to learn during your earthly lifetime. At the same time, because of the law of "cause and effect", your lessons may be tied to your previous choices. Of course, you can create more karmas (lessons) in a particular lifetime by your choices and actions. However, from a spiritual point of view, it is all well; it is joyful because the ultimate goal is to learn and evolve.

As stated previously, actions have consequences; therefore, sometimes, the soul, in concordance with God, His angels, and guides, elects some experiences to correct some actions taken in previous lifetimes or simply to learn what it is like to walk in the shoes of others. For instance, let's say that you wrongly killed someone's child in a past lifetime. You (your soul) can elect to come in this lifetime and experience the death of your child or someone may murder your child, or something of that nature. This is simply to experience what it feels like to lose a beloved child and to walk in the shoes of the child's parents. This is not a punishment in any way, as it was elected by you in agreement with God, your angels, or your guides. It is a choice that you have made prior to your current incarnation, and it is a lesson that you may have chosen to learn during your (current) lifetime. These types of lessons are referred to as karmas, but truly they are simply lessons and experiences based on past lifetimes.

Dharma or Experiences Elected before Birth

Do We Plan Our Lives before Birth?

Yes, we do. We plan our lives before birth because we have been given the gift of free will so that we can use it as we please. In fact, we plan many major experiences and events in our lives. At the same time, we have the free will to make decisions and shape our lives between these big events. Therefore, overall, we create our lives from start to finish, either with choices that we elect before birth or with those that we continue to make after birth. We choose the lessons that we want to learn during our earthly lives and also how we want to contribute to help others. These lessons can be a variety of things such as learning how to forgive, how to make money, how to love ourselves, how to develop resilience, etc. For instance, the life purpose of a beauty queen may be to learn the limitations of beauty. Therefore, if someone who is a beauty queen relies solely on her beauty, she may have experiences or realize later in her old age that beauty and the physical appearance have limitations and that there is more to life than physical beauty. This is just one example as these lessons can come in a variety of ways and flavors.

However, the ultimate lesson for all of us is to learn to love unconditionally. The lessons are somehow integrated into our life purposes. Everybody has distinctively designed lessons and life purposes that are uniquely theirs.

Do We Choose Our Parents before Birth?

Yes, we do choose our parents. When we choose our parents, we make agreements with them so that they will help us with what we need to learn, and they will give us exactly what we need in life. We choose parents who will be best suited for the things that we want to learn and for our life purposes. This may be surprising and difficult for us to comprehend, especially for those who had challenging times with their birth parents or who were abused by them. However, these are the experiences they chose for the growth of their soul, and they chose parents who would give them the experiences and lessons that their souls needed to learn.

Light of Truth

Here is an example of a true-life story but to preserve the anonymity of the person involved, I will call him James. James was born into a very dysfunctional family with an abusive father. Later, his father died, and James was placed in the foster care system in the United States. He went from one foster care family to another and had a very difficult and painful childhood. Long story short, James later turned his hardships into blessings, and today, he has changed the lives of more than 400,000 foster care children across the United States and around the world through his nonprofit organization, which is dedicated to helping foster care children. James succeeded in transforming his childhood hardships into blessings for others. His life's purpose probably was to bring blessings into the world and to create a brighter future for foster care and abuse children, and the only way that he could fulfill his life's purpose was to choose parents who would abuse and abandon him and to be later placed in the foster care system so that he could experience the lessons that his soul had chosen and his life purpose, which is to save and rescue abused children. James's goal was probably to develop compassion and understanding for these children and to rescue them. His soul was so compassionate and so loving that it was willing to choose parents who would give him the experience needed so that it could alleviate the suffering of millions of children in the world. This is how loving James (his soul) is. Yes, we do choose our parents, and we choose them purposely.

Oftentimes our parents are souls who we knew in other lifetimes. For people who have challenging times getting along with their parents, it is often karmic relationships and issues that they need to work on or simply parents who will lead them to the path of their life lessons and purpose; that's all. For those who are very close to their parents and have loving relationships with their parents, it is often souls who are "friend souls" or were deeply connected in an affectionate way in past lifetimes, and these parents are the right ones for the lessons and objectives of their souls.

Besides Choosing Our Own Parents, can You Give Us an Example of Something Else That We Choose before Birth?

We choose our names. We choose our first and middle names. Interesting, isn't it? You may ask, "How can that be possible when it is parents who choose the names of their children?" This is all part of the "grand illusion" of life. We simply make agreements with our parents on the names that we would love to be called during our adventures on Earth, and at the time of birth, our parents will simply give the names that they have agreed upon. This may seem too mysterious for us, especially when we see parents who are expecting children searching for the right names for them, reading books, and so on. This is all part of the pleasure of the game of life. In the end, no matter what name the parents finally decide to give their children, it will probably be the name that the souls of their children had chosen and agreed to have during their incarnations.

Do We Choose Our Physical Bodies or Appearance?

Yes, we do. We choose the race and gender that will best serve our life purpose; we also choose our physical appearance. Choosing our birth parents is a component of this, as souls often choose their physical bodies in a sense by choosing their parents. For example, an athlete may choose parents and a lineage so that their genetic makeup can support the area of sport that they desire to explore during their reincarnation. At the same time, they may choose parents who will be able to support their athletic pursuit and so on. Similarly, an athlete will be likely to choose parents who understand athletic values, who may be passionate about sport, or who are athletes themselves and will support or encourage them to get into a great athletic team during their lifetime. Coming back to the example of the beauty queen, the beauty queen may choose parents such that she will have an extremely beautiful physical body that will help her to win the "Miss United States Pageant" competition, for instance, so that she can have the experience of a beauty queen during her reincarnation.

We choose several details of our lives before birth, and at the same time, we have the free will to steer our lives in the direction that we want. However, when we incarnate, the ego or this part of our mind that operates from fear often pushes us to want things that are sometimes not aligned with our life purposes such as wanting to become famous, rich, or thin like top models, and so on. Also, because we may have thousands of incarnations in the third-dimensional plane, everyone will eventually play all the roles before they graduate so that they can have all the experiences and a broader perspective. Everybody will eventually play the role of the king, queen, president, poor, rich, homeless, celebrity, female, male, the prostitute, etc. The schoolroom of Earth is set up this way, and it is simply the reality of what it is. However, once we come here, because we forget everything from our past lives, people sometimes become angry with their life circumstances and wonder why they have the parents that they have, not knowing that they are the designer of their lives and had chosen their parents before birth. Consequently, many people want to be the rich and famous CEO, president, or celebrity. What kind of world would it be if everyone was a CEO, president, or celebrity? Who would check out our foods in grocery stores if no one wanted to play that role? Who would grow the food that we eat if no one wanted to be a farmer? Who would clean the office buildings where we work if no one wanted to be a housekeeper? The system is set up in a way that all the roles exist to make up a world that can sustain itself and support our lessons and growth. No life purpose is bigger or smaller than another; no life purpose is more or less valuable than another. It is an orchestra where all the players' roles are valuable, needed, and indispensable. Again, everyone will play all the roles at some point during their many incarnations.

The Truth about Life
CHAPTER 7

DEATH, LIFE AFTER DEATH, HEAVEN

"In a time of universal deceit, telling the truth is a revolutionary act."

–George Orwell

What Is Death?

Even the word "death" makes people scared. There is so much fear and suffering associated with death. We cling tightly to life and fear to die. Death is associated with the end of life, the end of everything, and so on. However, this is not the case. Death is simply a transition from the physical world into the nonphysical world. It is a simple transition, just as birth is the transition from the nonphysical into the physical world. There is no point in fearing death because life goes on after death. You existed, will continue to exist after your death, and probably will continue to be you and pursue your interests after your death. Of course, it is painful to lose loved ones, but no one truly dies. They just go on the other side and often look after you with love and compassion. Death can be the end of suffering for those who are in pain or have chronic diseases.

In fact, people find great peace and great joy when they transition into the nonphysical world. Please, there is no need to fear death in any way because death does not exist in spiritual truth; it is just an illusion. The best way to perceive death is to see it as "dropping the body" or "ejecting from the body"; that's all. There is so much sorrow, pain, and mourning for those who have passed away. Please, do not mourn over them because they are in a better place and well taken care of.

Death is also sometimes seen as punishment, and sometimes people are happy when their enemies die, or they even wish them dead. Also, there are times when people even kill others, falsely believing that they got rid of them. However, the person killed will just reappear after having left his body and be greeted with love by his guides and guardian angels and will be watching the killer rejoicing or trying to hide evidence of their actions. How crazy this is when you know that, in truth, death does not exist, nothing is hidden, and that life just goes on! Yes, the earth is a school, and it is a great school of illusion where people believe that death is the end of life and they fear it. Please rest in peace, knowing that nothing will ever hurt, tarnish, or extinguish your soul. You are eternal. You existed before this lifetime, and you will continue to exist throughout eternity. You are eternal.

What Happens When We Die?

When we die, we go back to heaven or the afterlife plane. Usually, we go through a tunnel of light, the purpose of which is to increase our vibration because heaven is a place of high vibration, a place where there is only love. Some who have had near-death experiences have reported seeing this tunnel of light or seeing a bright light.

After our death (transition), we also are greeted by our guardian angels who accompany us during the transition. The angels await and help this transition to be smooth. There is also a specialized angel, Archangel

Azrael, whose mission is to help people during this transitional passage to heaven. Death is simply a continuation of life in another form: the spiritual form. In reality, we never die because we are eternal light beings. Whatever God creates can never be extinguished. The best way to understand this is to see the body as a vehicle, and after life, we drop the body or the vehicle and continue our existence in another form: the nonphysical form.

Once in heaven, we go through a review of our lives with the help of our guides and guardian angels. It is a simple review to see what we have done and what we can improve on. There is no judgment, condemnation, or punishment of any kind. It is simply a review, as the earth life is a school; therefore, it makes sense that there is a review at the end of the incarnation and to have an assessment to see what we have done and how we can learn and grow from our experiences.

There is no judgment of any kind. There is indeed a time reserved to "review the life that one has just finished", which is synonymous with an evaluation or assessment. This assessment or review of our lives may be what is incorrectly translated as "judgment". This is good news indeed, and it shows the goodness and love of God. You do not even need to die before doing a review of your life with your guides, angels, and God. You can take some quiet time and assess your life to see what you can improve on and how you can become more loving.

I also want to mention here that at the end of your life, the only thing that will matter is how loving you were and how much love you gave and received during your earthly life. We are here to learn how to love. Therefore, during the review, you will be shown people with whom you connected in a loving way, people who you have helped, and how you positively impacted the lives of others. You will review your whole life, of course, but, at the end, you will only keep the love and lessons after the incarnation.

There are instances where you can be asked to come back to Earth and correct some of the choices that you made in order to learn from them. For instance, let's say that someone was a king during a lifetime, and he abused his power and unjustly taxed the villagers or farmers in his kingdom. He may be asked to return as a farmer or in another position where he will experience similar situations of injustice in order to learn the consequences of his actions. This may seem like a punishment, but, in truth, it is not; it is simply for learning purposes because, sometimes, the best way to learn is to walk in others' shoes. Keep in mind that it is all out of love and that the goal is to help people learn. This is what some people refer to as "karma", but, again, they are simply learning experiences.

Because the experience on Earth can be sometimes very challenging and traumatic, some people also go through a healing phase after death and a period of transition to release anything that they may have collected during their travels on Earth.

We stay in heaven until we feel like it is time to reincarnate again. We make the decision of when we want to come back and what we want to learn or work on.

Does Heaven Exist?

Yes, but heaven is more about vibration. Heaven is not somewhere far away, neither is it somewhere in the clouds. Heaven is right here among us. Remember that we have guardian angels around us all the time, even though you may not see or hear them. There are also deceased loved ones who watch over us as well. It is a matter of vibration. Everything that is heavenly is right here. Yes, there is an afterlife plane that we go to after death; it is a plane of higher vibration as well. God is in our everyday reality. God is in every cell of your body and everywhere.

Where Do We Go When We Die?

We all go to heaven. Everyone goes to heaven, no matter what they have done or not done in life. This is because there is nowhere else to go. I will add that even the cruelest dictators go to heaven—even though this may be shocking to some. However, when we honestly think about it, we can see that this shows the grace and mercy of God and only confirms how unconditionally loving He indeed is. This is good news, isn't it?

What Do We Do in Heaven?

We simply continue to learn and grow. Those who are artists or musicians, for instance, continue polishing their artistic skills, composing songs, learning new dance moves, etc. People also collaborate with others there to learn and evolve. We also take classes in what we are learning. There are also classes to learn how to love. Those who are healers (physicians, nurses, physical therapists, massage therapists, yoga teachers, etc.) continue to learn about healing and various healing modalities. Those who are teachers or inspirational speakers may work on polishing their speaking talents and gifts. In fact, we are free to do whatever we want there. Some souls also assist people who are on Earth by giving them inspiration for their artistic work. For instance, some famous artists who have accomplished great things on Earth continue their work in the heavenly plane and assist musicians and other artists on Earth by sending them information that they receive as inspiration—sometimes in their dreams or in various ways. Some of the souls assisting others have already graduated from the third dimension and don't need to come back, while others still need to come back to learn. Nevertheless, souls continue their journeys and works in heaven.

There are infinite possibilities of what we can do in heaven, and God has given us the free will to do or be whatever we choose. We also take time to prepare for our next incarnations after we go through the review phase—if we are ready to come back and incarnate again. We take time to plan our future incarnations or lifetimes with the help of our angels and guides.

What Is Heaven Like?

It is beautiful beyond what we can conceive and describe with words. It is a place of unconditional love, peace, and blissful joy. It is filled with light and love. There are many angels and guides that help us there. It is truly a paradise, a heaven indeed, just as its name implies (here, I am referring to the afterlife plane). At the same time, the heavenly beings such as our assigned guardian angels are constantly around and with us here on Earth.

The Truth about Life
CHAPTER 8

SOUL AGE, SOUL GROUP, SOULMATES

"The truth is always the strongest argument. Sophocles Truth is a thing immortal and perpetual, and it gives to us beauty that fades not away in time."

–Frederick The Great

What Is Soul Age and What Can You Tell Us about It?

In spiritual truth, a soul does not have an age; a soul is ageless. However, some people have a greater awareness in spiritual maturity than others because they are further in their spiritual evolution than others, and some refer to them as "old souls". Old souls are souls that have reincarnated a countless number of times; have come far along their spiritual evolution; and have learned valuable spiritual lessons, accumulated knowledge, and have more awareness of love. They are very compassionate, loving, and more inclined to opt for unity, accepting others as they are, even those who look different from them. On the other hand, "young souls" have had fewer incarnations on a third-dimensional plane such as the earth and have less awareness of love and compassion. Therefore, young souls are more likely to live in the ego's world of anger, hatred, discrimination, war, etc. However, I would like to reiterate here that a soul is truly ageless, and the terms "old souls and young souls" are simply attributes to describe those who have more awareness regarding the truth about life and love in comparison to those who have a little less awareness on that regard. However, I believe that we all have the same potential in spiritual truth and are all capable of love, no matter our soul ages.

The bigger the difference between the soul ages of two people, the more difficult it will be for them to get along because they will simply have challenging times to see things with the same eyes and to have the same perspectives, but that is OK; it's meant to be that way. The school of earth is designed in such a way that everybody will have the opportunity to interact with those who have different conditioning or different levels of awareness. Conditioning refers to a collection of things that are not intrinsic to our divine nature and that seem to separate us from one another. Examples of conditionings are: education, social background, race, environment, place of birth, beliefs, religious backgrounds, childhood experiences, ideas collected from the media, ideas learned from parents or parental figures, soul ages, and so on. The bigger the differences in conditioning two people have, the further apart they will be in their ability to get along and accept each other. Among the various elements that can separate two people, the difference in their soul ages is one of the main factors that play an important role in the ability of two people to get along or to feel affinity with each other.

The school of Earth is orchestrated in such a way that everybody will have people amongst their family, friends, coworkers, relatives, etc. that will be further apart from them in terms of their soul ages. This gives better opportunities for learning and growing in our abilities to love others unconditionally. Otherwise, how good would it be if we were only to love those who were like us? It would be too easy, wouldn't it? "If you love only those who love you, what reward is there for that? Even corrupt tax collectors do that much" (Matthew 5:46). Also, being only among those of the same ages and who have similar interests as you will defile the very purpose of the schoolroom of Earth, which is learning to love all people unconditionally, including those who are different from you, which is one of the reasons why we all are here.

What Is a Soul Group?

A "soul group" is simply a term used to refer to souls that often incarnate together. We connect and collect friends throughout our various incarnations. Some souls are so connected and love each other so much that they often plan to incarnate together. For instance, some always come as wife and husband to support each other in their learning. Others often come as brothers or sisters. In fact, we are truly free to do whatever we want, and it is normal that some souls have more affinity with each other and want to remain together. Oftentimes, when they reincarnate and meet for the first time, they have the feeling that they have known each other for a long time, and this feeling is often true because, indeed, they did know each other in previous lifetimes, and that is why they have this feeling of déjà vu or connectivity.

To summarize, a soul group comprises souls who we knew in previous lifetimes, who incarnate with us and who are more likely to be in the circle of people that are in our lives. They may be in other parts of the world, other cities, other states, other countries, or other continents, but they are people with whom you will be in contact frequently.

What Is a Soulmate?

When we think about soulmates, we often think about love relationships, but, in truth, soulmates are souls that we knew in previous incarnations; they can be our love partners but also our parents, children, brothers, friends, work partners, etc. whom we knew in previous lifetimes and have reincarnated with us. For that matter, we have several soulmates even in love relationships because we have probably played the roles of spouse with different souls in past lifetimes.

I would like to mention here that, sometimes, we can also have karmic relationships with people, and these people will show up in our lives as spouses or love partners. Therefore, the first time we meet them, we feel attracted and connected to them and thought that they were our soulmate, only to later discover that the relationship turns sour.

The Truth about Life

CHAPTER 9

RELIGION

"Half-truth is often a great lie." –Benjamin Franklin

Religion
Subchapter 1

THE TRUTH ABOUT RELIGION

What Is the Truth about the Various Religions?

When the dharma has been fully realized, i.e. when all of the previous lifetimes have been completed, when all the learning, lessons, and growth are fully realized in the human experience, the human is considered "enlightened". Since the beginning of humanity, only a few human beings after reaching this stage of their evolution, after becoming "enlightened", no longer need to come on Earth yet deliberately choose to come back to Earth to help guide and alleviate the suffering of human beings. There are only a handful of human beings that after attaining this stage of their evolution have made this honorable decision of service, to name a few: Jesus, the Buddha, Moses, Yogananda, the Saints, Saint Augustine, Saint Germain, and others. These are only names of human beings, but, in truth, they are vehicles for Light to shine through. They chose to be volunteers of service, ambassadors of Light; they chose to come and serve humanity, to help, to teach, and to conduct a specific mission that will help and propel humanity in its evolution.

THE TRUTH ABOUT LIFE

After the enlightened beings such as Jesus and the Buddha had served and shown the way to Light, the way to Love, a religion was created in their names. Human beings enjoy the idea of religion, the storytelling and the feeling of security and hope that it provides. I will say that the truth is simple; it is humans themselves who created a preplanned agenda on how to know God or the path to God that involves tools, rules, formulas, and practices that they call 'religion'. All of these are human activities. Above all of these human activities, there is truth, and the truth is simple: The path is love. The only one religion that one can adopt is love.

The truth is that God has sent several people to help us on Earth, and their messages got turned into religions. For instance, Jesus was not a Christian when He was on Earth, and His intention was not to start a religion. He came to teach about love, about forgiveness, and to do God's work by showing the way to love. After His death, a religion was created in His name to designate his followers as "Christians", which means "Christ followers" or those who adopted his teachings. Thus, Christianity was born. This was created after He left. All religions were created by humans and came from the mind of humans.

This was the same case for Buddhism as well, which was created after the Buddha to label or designate those who were following His teachings. It is God who sent Buddha as well, or, to be more precise, He chose of His own accord to come to serve as a messenger and ambassador of Light, of Love. Indeed, the Buddha came to teach about compassion, which is a form of love, coupled with understanding. His mission was to bring peace, compassion, and acceptance in a part of Earth at a time when there was a caste system, great division, and great suffering. Compassion, love, acceptance, forgiveness, and kindness were the message of His service and teaching.

Now, regarding which religion to adopt, I will say that if one wishes to adopt a religion, let them adopt the religion of love, compassion, acceptance, forgiveness, and kindness for all.

The Blind Men and the elephant

A retelling of a famous Indian fable

A long time ago, there lived a group of old men in a faraway village in India. These nice and loving old men were born blind; they could not see and had to imagine the wonder and all the beauty of the world around them. The villagers loved the old men and kept them away from harm. One day, the old men heard rumors from the villagers that a strange animal called an "elephant" was in their village. They had heard so many stories about the elephant and had heard that elephants could trample a whole forest, carry huge burdens, make unimaginably loud noises, and also carry princesses on their backs to travel, and so on.

The old men become fascinated about this strange animal called "elephant" and all they had heard about it. They discussed day and night about what the elephant might look like. One night, in their discussion about this strange animal, they argued among themselves:

One old man claimed, "An elephant must be very powerful and gigantic. I have heard stories that elephants can trample whole forests and build roads."

"You are wrong," declared another old man. "An elephant must be a gentle and graceful creature. I have heard that elephants carry princesses on their back and gently rock them while they are traveling on their backs."

"You are all mistaken," stated another blind man. "I have heard that elephants can pierce men's hearts with their ugly horns."

Another blind man interjected, "Elephants do not exist at all. There cannot be such creatures that are able to do all these things and fit all the descriptions and stories that people are telling in the village. We all know how people can exaggerate and invent stories in our village. We are all victims of a cruel joke."

In the midst of the disagreements, and tired of the arguments among themselves, the blind men decided out of curiosity to go and check out by themselves what this strange animal that everyone was talking about in the village looked like and to get to the bottom of the truth once and for all. A young boy from the village came and led them to the center of the village where the elephant was kept so that the old men could inspect and touch it to satisfy their curiosity and cease the arguments between them. One by one, the old blind men started to inspect and touch the elephant with their hands.

The first blind man reached out and touched the trunk of the elephant and said, "An elephant is like a snake."

The second blind man reached the leg of the elephant and said, "An elephant is like tree-trunk."

Another one reached the animal's ear and said, "An elephant is like giant fan."

Another old man reached and touched the side of the huge animal and said, "You are all wrong. An elephant is like a solid wall."

They started to fight among themselves because they were all convinced of their experiences, and each one was sure that they were right and the others were wrong since they had touched the animal with their own hands and were confident in what they had touched.

The fable of "the old blind men and the elephant" is a great illustrative example that perfectly reflects the state of the various religions and the mindset of many of people when it comes down to the religions that we currently have on Earth. Like in the case of the "old blind men and the elephant" there are many stories being told about others' religions. In this fable, we can see that there were stories and rumors stating that elephants could build roads or pierce men's hearts with their ugly horns, etc., which were untrue and simply exaggerated stories and false rumors. Similarly, in our world today, many people create stories about other religions that are untrue and believe these stories without even reading a single word of the scriptures of these religions. For instance, some people believe that meditation practice in Buddhism is evil, while, in truth, meditation is just a word or label given to the practice of quieting the mind to help get us out of the egoic thought screen. What can be evil in following your breath, practicing silence, or listening to soothing music? There are even scientific studies that have proven the benefits of mediating. Yet, many people hold such a belief without even understanding what mediation is or without even trying once. Once again, like the blind men who hadn't yet touched the elephant, we believe we know what an elephant is and start arguing and fighting among ourselves.

Even after touching the elephant, i.e. even after adopting a religion, we still really don't know what an elephant truly is. Indeed, we are in a similar situation as the "the blind men and the elephant" as, currently, every religion is convinced that they know "the truth and the only truth" that exists while they may only have a partial truth. However, a partial truth is not the whole truth because the reality is much bigger and grander than what we can perceive or understand.

In reality, there is a little bit of truth in many religions, and yet none of them holds the whole truth. For instance, some religious practices claim that reincarnation exists, and it is true we do reincarnate. Yet, there are other misconceptions associated with the idea of reincarnation in these religions such as perceiving karmas as punishment or being forced to reincarnate as an animal if someone did not live a pure life, etc. However, the truth is that there is no punishment, and karmas are simply lessons, and we do come back in human form regardless of what we have done in previous lives.

Other religious practices claim that the earth will not come to an end and that heaven will be right here on Earth. There are some truths in that as well because there will not be an end of the world, and the earth is simply going to shift into a fourth and fifth-dimension vibration, which will be more a peaceful and heavenly-like place than it currently is. Yet, this is a partial truth as well because heaven does exist, and it is the afterlife plane where we go when we die. Again, partial truths are not the truth, but they seem very convincing because they have a tiny truth in them, which makes them very believable.

Other religions claim that Jesus is the Son of God and that God sent Him to help humanity. This is true as well. Yes, God did send Him to help humanity, but He was not the only heavenly being sent by God. Currently, there are millions of highly evolved heavenly beings that God has sent to help the earth during this Earth transition phase. These beings come from the fourth, fifth, seventh, and even twelfth dimensions. Some incarnate as humans and are like ordinary people. They are walking on Earth right now among us. Many of these beings are lost, while a small percentage of them have woken up from the illusion and are carrying out their missions of helping to raise the vibration of our planet by working as healers, philanthropists, teachers, missionaries, and so on.

Some religions swear by the practice of fasting, and it is true that fasting does have a cleansing effect, helping remove chemicals in the body, and thus helping to increase vibration. The practice of fasting was introduced probably by heavenly messengers, and it is an ancient spiritual practice that is really helpful. Therefore, there is some truth in that as well, but, again, a partial truth is not the whole truth.

To summarize, like the story of "the old blind men and the elephant" every religion holds some form truth, a partial truth. Yet, none of them holds the whole truth. If someone wants to adopt a religion or spiritual practice, then Love is the religion to adopt. Learning to love and loving all people unconditionally is the only truest form of religion, and we all are well capable of loving others, no matter who we are and regardless of whether we are religious, nonreligious, atheists, or do not care about religion at all.

Religion
Subchapter 2

GOD

Who or What Is God?

We have been told so many things about who or what God is. We have been told that God is like a wise old man who resides far away somewhere in the clouds in heaven. We have been told that God is like a king sitting on His throne in heaven and is worshiped by angels and the righteous people in heaven. We have been told that in His kingdom, the Kingdom of heaven, God rules with justice and a strong hand. We have been told that God is like a judge who judges and will judge us at the end of our lives, sorting out and separating those who followed His commandments from those who behaved badly on Earth. We have been told that on His heavenly throne, God is surrounded by angels in long, white dresses who worship Him and sing choruses to Him. We have been told that from His heavenly throne, God looks at the earth, screening and watching all that we do. We believe that God is a male and not a female because we have been told that God is a father, not a mother, and that God is a king, not a queen. Imagining God to be a queen is very shocking for us because of the way we have been programmed and all of the things that we have been told; this very idea of imagining God to be of a female gender is really shocking for many and can even be considered by some as blasphemy.

All these ideas and things that we have been told are not true. So, what or who is God? Although it can be difficult for us to fully comprehend the concept of God, here are some examples and analogies that can help to understand "That" which we call God:

God is Energy, and so are we. God is just a word that is representative of an Energy field. God is Light and Sound, and so are we. Isn't it interesting that some spiritual scriptures refer to God as "the Father of Light"? "Every good gift and every perfect gift is from above, coming down from the Father of Lights …" –James 1:17

God is pure Light and Love. To have a better understanding of what God truly is, you can see God as Energy, Light waves with intelligence, or you may also see God as a living Light that is fully alive and has intelligence of its own. What an amazing Intelligence! Look around and see for yourself; look at the complexity involved in the cellular and molecular biology of the human body. Look at the complexity and all the mathematic equations, physics, calculation, and all the science behind the creation of a galaxy, and you will come to understand that there is truly no word in human terms to describe this "Infinite Intelligence" that we call God.

God is All-in-All. What does this mean? It means that God is in everything and God is everywhere. The whole Universe constitutes God, is made of God by God and is God. Another analogy is to see God as the fire, and everyone and everything is like a sparkle of that fire of God. We are extensions of God on Earth, and so are the trees, the seas, the animals... Because God is All-in-All, we often hear people say, "God is omnipresent." Many of us know that God is omnipresent but do not fully understand that it is because God is everything and God is in everything.

God is the clay out of which we were all created. Take clay, for instance. A lot of things can be made out of clay. God is the clay, and we are the products or creations of this clay. From clay, you can make different forms and shapes, e.g., a male, a female, a tree, a dog, etc. All of these forms may seem different in appearance, and yet, at their very cores, they are made of the same material—clay (God). The same applies to us as well. We seem different only in appearance, but the very essence of who we are is God. We are made of the same substance and material of God. We are God. Let's say that there is no concrete word to describe who or what God is, as God can take any form He wishes, and we are all God and connected to God. God is in us, and we are all parts of God.

God is the Maker and Player of life. He plays all the roles. In this analogy, life can be compared to a video game, with God being both the creator of the video game and the one playing all the roles in that video game. When you play a video game, you can become so absorbed and caught up in the game that you may think that you are the game character. This is what happens with life as well. When we incarnate on Earth, we become so caught up in the "game of life" that we think and believe that we are the characters in the game. This is an illusion, of course, but the illusion of life seems so real that we are completely immersed in it and believe it. "Life" is like a "dream", an "illusion"; the experiences of the dream seem very realistic, but once awake, you realize that it was just a dream. The only way to fully realize that life is nothing but an illusionary dream that lasts only for a time is to fully wake up from the illusion and truly know the truth about life.

God is 100% pure unconditional Love. God is 100% goodness. God is 100% compassion. This is the true essence of what or who He is. Contrary to our beliefs, God does not have ego; He is pure unconditional Love. God does not have human characteristics of the ego such as anger, resentment, or all other human fear-based behaviors or characteristics.

God is our Father, our Creator, and God created us all. Indeed, the Energy field that we call God created the whole Universe including us. Instead of imagining God creating us out of some material, it is best to see God as projecting or multiplying/dividing Himself countless times to make up "us" and the whole Universe. That is why we are part of God; He is part of us, and we are Him. That is why we can be seen as offspring of God or children of God. God created the body or vessel that we use on Earth to experience the physical life; He multiplied/divided Himself countless times to make us (our souls) and the Universe. With that in mind, the big bang theory is not totally false and not totally true either since God projected Himself like "splitting" Himself to make up the Universe. At the same time, the creativity theory is not totally false either.

God does not have a gender. God is neither male nor female; God is both male and female at the same time. That is why the closest attribute of God sometimes is Father-Mother-God. In fact, God has both male and female energy, and so do we. For practical reasons, I will refer to God in the subsequent chapters of this book as a male (He, Him, etc.), but, in truth, the best way to refer to God is "He / She" or "It" because God does not have a gender or God has both genders.

Clarification of Some Misconceptions about God

Is God a God of Vengeance and a Jealous God?

No. God is 100% unconditional Love and completely egoless. Therefore, God does not get angry. He never judges and is all patience. No matter the mistakes or what we do in life, God never condemns or judges us. Love is the essence of who or what God is. Anything about Him that is not aligned with Love and does not come from Love, such as God being angry, judging, punishing, being jealous, and so on, is not the truth about God.

God never said He is a God of vengeance or a jealous God. These statements and all other similar statements are lies collected over millenniums to bring fear and to control humans on Earth, and, unfortunately, they have been reported in some scriptures and attributed to God. I would not go further into detail on this but I will simply say that anything that does not resonate with Love does not come from God and is not true. No matter the religious affiliations or beliefs, anything that is contrary to Love is not the truth about God and not from God.

Here are some derivatives of Love: compassion, forgiveness, kindness, patience, gentleness, acceptance of others, equality (we are all equal in the eyes of God), Oneness (we are all One and connected to each other), peace, service, helping others, unity, supportiveness, sharing, loving everyone (even your enemies) … and so on. Anything that goes against Love is not of God and does not come from God.

Does God Punish People?

No. God does not punish anyone. God is not a God of punishment. Of course, there are consequences in life, and life on Earth is created based on some spiritual, physical, and metaphysical laws. The law of "cause and effect" is one of these laws. Therefore, every action has a consequence. When you put positivity out, positivity will come back to you. The same is true for negativity as well: when you put negativity out, you will reap the fruits of your actions; it is not God who is punishing people, these are simply consequences of their own actions. It is the law of "cause and effect". In that way, life is self-correcting and fair in a comprehensive term. In the realm of God, mistakes do not require punishment but simply correction. Notice here that "cause and effect" is very different from "punishment".

Does God Sometimes Test People?

No. God never tests us and does not desire to test anyone. God has nothing to prove to anyone, and God is not avid of our love, worship, or anything of that nature. God is God. It is not the essence of who God is, and anything related to "God testing us" to prove something to Him is not truth. There is no test. There are of course challenges in life, but they are not tests from God. The challenges are simply experiences that we can learn and grow from.

Truth Detecting Tips

When I talk about truth, I am referring to the spiritual truth, which is the ultimate truth and the only truth that exists since anything else is only an illusionary dream and false.

How do you know if something is true or false? When something is false, it creates fear, contraction in your body and uneasiness. On the other hand, something that is true creates peace, relaxation, joy, hope, and appeases. It is something that you will have to discover; you need to train yourself to recognize the truth in the midst of lies and distortions.

Let's use a concrete example to illustrate how to distinguish between something that is true and something that is false: If someone tells you that you are ugly or you say to yourself that you are ugly, it will contract you and make you uneasy, frustrated, or even angry. Why? Because it is not true: In spiritual truth, you are a magnificent beautiful being of Light and a shining child of God. This is who your true essence or your soul is, and the "true you" is your soul, not your body, which is just a temporary physical vehicle. The "true you" or "your true self" or "your soul" is really beautiful indeed, and some people who are clairvoyant can see the beauty of your soul shining through. In fact, if you can only see yourself through God's eyes, you will know how beautiful you are indeed.

Here is another example: When someone tells you that at the end of your life, you are going to be judged and that based on that judgment, you may go to hell, it will create a deep fear and contraction in your body. It will scare you. Why? Because it is not true. The fear that you feel is an indication that the idea of punishment and judgment is false.

Religion
Subchapter 3

JESUS

Is Jesus the Son of God?

Yes, Jesus is a Son of God as we all are.

Is Jesus the Only Son of God?

No, we are all children of God. The truth is that we are all children of God and equally loved, equally valued, with the same potential in spiritual truth. I believe that the same spiritual scripture that mentions that Jesus is "the only" son of God also mentions that we are children of God. However, we cannot all be children of God and Jesus be the only Son of God at the same time. One of these statements must be false since they are contradictory. The truth is that we are all children of God.

Jesus is just among those beings who are more advanced in the evolution process, and He came to teach, heal, and enact God's plan of peace on Earth. It would be more accurate to see Him as a big brother rather than the only Son of God. In fact, these highly evolved beings such as Jesus, Buddha, Moses, etc., like to see themselves as our brothers and not above or superior to us, as we are all children of God and equal in spiritual truth.

To better understand this, we can use the analogy of school and can perceive these beings, such as Jesus or Buddha, as being in middle school in the evolution process, while we on Earth are in first grade. A middle-school student is not more valuable or superior to a first-grade student. A middle-school student is just more advanced in his or her education; that's all. Eventually, the first-grade student will get into middle school one day. No loving father will see or say that his first-grade children are less valuable than his middle-school child or that his middle-school child is his "only son" just because he is in middle school. Fortunately, God is a wise and loving parent and sees all of His children as equal with the same value and potentiality, whether they are in first grade or middle school. He loves us equally. Eventually, every one of us will get to middle school one day, even if it will take thousands of incarnations, millions of years, or eons.

I would like to add here that Jesus had many other incarnations, and the incarnation when He was known as Jesus was the most prominent and His last one, even though He is still well alive and helping us today. He was once known as the prophet Ezekiel described in the Bible in a previous incarnation. He had several other incarnations unknown to us.

What Is the Meaning of the Death of Jesus on the Cross?

Jesus is among the few beings who after completing their dharma on the earthly plane and after becoming "enlightened" decided to come back to serve and help guide humanity. Like every human being, the dharma is set before birth. In fact, before the physical experience (life on Earth), there are choices to be made; there is an agreement that every human signs, saying, "Yes, I will take on this experience. Yes, I will serve in this way or that way." The being known as Jesus had chosen and agreed Himself before His birth that He would sacrifice so that others may be free. In His service, He decided to be an ambassador of Light, a way shower, and to be a vessel for the Light to shine through. The Light that we are talking about is inconceivable, and that is where the focus needs to be when we are talking about Jesus and the sacrificial act of being nailed to the cross. Far too much attention and storytelling has been given to this act of being nailed on the cross. Humans have created stories about this. People have even written books about this. The act of Jesus dying on the cross has become imbued; it has become somehow "more" and even glorified sometimes. I am not saying that it is insignificant, but I only speak about the pure truth about the Divine Light that operates through each human, including the "Jesus". That Light was not on the cross, even though the body was placed on a cross. In truth, the Light could never be sacrificed; that does not exist. The body can be destroyed, scarified, and tortured, but the Light cannot. The Light carries on throughout eternity. The Light is indestructible; it is eternal.

The true meaning of the sacrifice of Jesus on the cross is to serve as "a model" and to give to humans hope. In fact, the human mind needs images and ideas such as hope. Hope is an idea and the image of Jesus on the cross is now the cornerstone to Christianity. It provides hope and healing. It provides incentive for humans to reach beyond their limitations, to see that this being, Jesus, devoted His life to Light and at the end gave a sacrifice of Himself. He made this sacrifice so that human beings can feel supported. Jesus' example was set this way so that humans could benefit. Human beings need to feel hope. Human beings, when they see the image of the cross, say, "Oh, Christ has offered His body so that all of the wounds of humanity could be washed cleaned and He did it." To sacrifice oneself to alleviate the suffering of others, can you imagine it? No, most human beings cannot imagine it. Imagine having your hands pierced through hammering and with nails and to look at the men doing this, to look into their eyes and to forgive them instantly: As they hammered, offering forgiveness to them in the moment? This is invincible compassion! It exceeds the physicality; it exceeds everything that we humans have been taught and seen so far.

This example of sacrifice on the cross is helping human beings to learn invincible compassion, to preserve hope and to have belief and faith beyond the physical body, beyond the suffering. Jesus, the energy of Christ Consciousness, selected and volunteered to be the "ambassador" and the "symbol". Human beings need these symbols and it will carry on for the legacy of Christ and for many generations yet to come. That is how powerful the symbolic act of the sacrifice of Jesus on the cross is viewed and the energy of it: Christ Consciousness. It is powerful.

Did Jesus Resurrect from Death?

Yes, he did and every human resurrects. All of us have many previous lifetimes behind us and if you are still here today reading these words, this means that you have resurrected somehow. You have been reborn; therefore, in a way, we can say that you have resurrected somehow.

Once again, do you see how humans have created stories to glorify something, a process that happens with every human? Every human resurrects. At the point of exiting the body, the soul rises up.

Religions create thoughts and stories in the mind for humans to give them hope, faith, and encouragement. This is neither good nor bad; it is just an occurrence; it is a phenomenon. That is what we call religion, and this is what religions do: they create stories; they magnify, embellish, glorify, and make things extraordinary to give people hope and faith.

Yes, at the point of leaving the body on the cross, a resurrection occurred, and then reappearance. Many are said to have seen Jesus in His physical form. This is quite possible, though very rare, extremely rare and it is written, "I saw the Jesus. I saw His form after the crucifixion." Yes, it is so. The physical form reappeared. Oh, yes, Jesus reappeared in His physical form after His death, and many men, women, children, and animals witnessed it. Now people have become so fascinated by the story of Jesus that researches have been conducted to study this and they will continue to study it.

Did Jesus Walk on Water?

Yes, He did.

Is Jesus Coming Back?

No, Jesus is not going to come back in the way that we are expecting—in a cloud, at the sound of the trumpet, in a spectacular way, and so on. The Book of Revelation was written in a parabolic way; it is a metaphor. No, Jesus will not physically come back. I will explain later in this book some of the metaphors used in the Book of Revelation and their true meanings. However, there is no need for Jesus to come back that way; it is useless since He is already here with us. He is everywhere. Indeed, Jesus is currently helping people on Earth, and He operates on a spiritual plane. He helps all who call on Him, no matter their religious background or beliefs. He is omnipresent like God and can help many people at the same time. He is one of the main beings who watch over this planet and help us in our evolution. Straining on whether Jesus is coming back or not is not necessary because He is already here in our everyday life and working unceasingly for our welfare. This is good news indeed because, on His heavenly plane, He can help countless of people at the same time, and I believe it is better that way.

Religion
Subchapter 4

ISLAM AND MOHAMED

What Can You Tell Us about Mohamed and the Birth of the Islamic Religion?

Islam is another iconic religion. Mohamed is a sacred being; he was the keeper of keys. There is a prism light through Islam that is tied to the Holy Cities. This light can be felt upon a bow, in the prayer upon the third eye. The prism is like a spinning diamond and Mohamed was holding and upholding this light. Islam itself is a wonderful clean and pure religion. Human beings enjoy the ideas of religions and the storytelling and the feeling of security and hope that they provide. Religions are here to direct and help human beings; these religions are like bridges but they are not and should not be considered as "The Source". The Source, the Light, is something that we cannot speak of. It is beyond language, it is inconceivable and it is that which conducts all of these religions. Although human beings will have wars and discussions about this, saying this religion is right, this one is wrong;

this is white, this is black, and so on, above all of these, the Light presides. The Light is on and it shines through every religion.

Did Islam Originate from the Pure Light and Love of God?

Yes, all the religions do.

How Has Islam Become What It Is Today? Why the killing? It is Completely out of Love.

This is because of humans' behaviors and human tendencies. It has gone mad. The human mind has gone mad and has got hold of the religion. There is an epidemic that creates human suffering: it is the egoic mind and when it gets out of control, madness happens. However, that has nothing to do with the religion itself, which, as said, is pure and clean; it is of a prism and through the third eye, through the bow, it is a practice in Islam, the light shines through; it is a beautiful star. The religion itself is separated and far removed from human interpretations and humans' madness. On the whole, the entire planet is now under this siege of this epidemic; the epidemic is the egoic mind. When the egoic mind intercepts upon the religion, madness happens. This is why the killing, this is why the interpretations have bent and construed the rules of the religion and the beliefs of the religion. The religion has been tainted with human brain activity. However, in its purest form, Islam, it speaks love. It is a symbol for love.

Every single religion is very useful. It points you to Light, but not every religion is Light and it will depend on every human being, how they will interpret and how they will create a relationship with the religion. Many humans confuse the religion as "The Source" and that is not so.

THE TRUTH ABOUT LIFE

If you are quiet, in silence, and you allow the brain to rest, and you feel something else, something Higher, something far greater than the religions and the doctrines, this "something" is the Light.

Islam, like the other religions, in its original form is a clean and pure religion that speaks about love and that came to introduce some spiritual practices such as prayer, bowing, fasting, and to teach about love. As with the other religions, it has somehow become tainted, changed, and intercepted with human interpretations, additions, editing, and human add-on untruths and misconceptions. These distortions, misinformation, and untruths led to what we have now on Earth with people killing others to convert them to Islam, to wars called "holy wars" and so on. How can a war be holy anyway? How can forcing other people to convert them against their will to another religion come from God, come from Light? This is happening because humans have got hold of the original message of Islam, which was a message of love and unity, and have transformed it, which has led to all the madness that we are seeking today in the world.

Keep in mind that this happens with other religions as well. Similarly, the ideas of God being a jealous God and God of punishment have been introduced by humans themselves and added to Christian scriptures leading to the confusion and madness as well. Also, keep in mind that the madness that we are talking about is not limited to Islam only but touches all religions. There have been wars on this planet about Christianity between different branches of Christianity. There have been people killing others in the name of Jesus while the original message of Jesus was about love, compassion, and forgiveness. All this madness clearly illustrates what can happen when the human mind (the ego) takes something that speaks about love, compassion, and kindness and turns it 360 degrees into something completely out of love. Instead of judging each other, and condemning other religions, we should know the truth that all of these religions in their original forms speak about love and that there is no bad or wrong religion. The key is to know and recognize the truth that sometimes in the history of humanity, people hungry for power and control, or for whatever reason, have deviated from the original messages of religions, added on, taken out, and created something coming from their own egoic minds and their own misinterpretations and untrue concepts.

Truth Detecting Tips

No matter what religion or belief you have, or even if you don't believe in anything, every time you hear anything that does not talk about love, unity, and oneness such as vengeance, punishment, an eye for an eye and a tooth for a tooth, forcing others to convert them, God has ordered his people to kill others, we are the chosen people and they are the outcast ones, ours is better and theirs is wrong, division, women are as less than men, children are less than adults, anything that will go against others' free will… etc., know that these come from the distortions and added messages from humans themselves and that they are not true, do not come from Light, and do not come from God.

Other Truth Detecting Tips:

Inconsistencies:

God never changes and then God changes His mind.
Jesus is the only Son of God and then we are all children of God.
God is fair and then God has His favorite people "the chosen people".
God is unconditional love and then God is a jealous God.

Fear-inducing Statements and Actions

Punishments
Wars in the name of God

Holy wars
The wrath of God
The anger of God

Dictatorial Statements and Actions

God has commanded.
God has ordered.
God has decreed.
God has ordained.
God commanded them to kill everyone and to take over their land and all the possessions of the others.
God told them to kill everyone who would not respect this or that.

Nonsense and Insane Promises

God promised them to take the lands and wives of others.
God promised them virgins in heaven if they followed His commandments.

Religion
Subchapter 5

PRAYER

What Can You Tell Us about Prayers?

The topic of prayer is an interesting one; many people debate its purpose and some do not believe in it at all. The truth is that prayers work and do get answered. Every prayer is answered. However, how prayers are answered by the Divine follows some essential rules and principles, which I will refer to here as the "Seven Essential Keys of Prayers." Knowing these keys is crucial. Here are some essential keys to how prayers work.

Prayer Essential key #1: the law of free will

Because of the free will law, Universe/Source/God will not intervene unless we ask for help and assistance. This is why prayer is important because it allows the Universe to help you. Indeed, God, the angels, and our guides are happy when we pray because it opens the doorways for them to intervene. Prayers are like sweet music in their ears. When you are praying, it is as though you are ringing a bell saying, "I am here. I am giving you my free will 'pass' to intervene in my life regarding this matter or problem." Thus, Source will intervene and help you.

Prayer Essential key #2: sincerity

When it comes to prayers, sincerity is everything. Sincere prayers from the heart are always answered and will yield the solutions that one is seeking. On the other hand, saying memorized prayers and reciting or rehearsing them like a robot without paying great attention to the prayers will not yield any result. This is because it is the sincerity, the intention, the honesty, and the faith behind that prayer that matters the most and it is what God responds to.

What do I mean by sincerity? I will use two popular prayers to illustrate what I mean by the importance of sincerity in prayer. The first prayer is in the illustrative examples is the "The Lord's Prayer" and the second prayer is "The Hail Mary Prayer".

The Lord's Prayer: *"Our Father in heaven, hallowed be your name. Your kingdom come, your will be done, on Earth as it is in heaven. Give us this day our daily bread, and forgive us our debts, as we also have forgiven our debtors. And lead us not into temptation, but deliver us from evil."*

The Hail Mary Prayer: *"Hail Mary, full of grace, the Lord is with thee; blessed art thou amongst women, and blessed is the fruit of thy womb, Jesus. Holy Mary, Mother of God, pray for us sinners, now and at the hour of our death. Amen."*

THE TRUTH ABOUT LIFE

<u>Example #1:</u>

Let's say that someone has learned and memorized "The Lord's Prayer" or "The Hail Mary Prayer" and his prayer practices consist of reciting these prayers a few times daily without great sincerity or not even paying great attention to what he is praying or being distracted by something else while he is praying because he says these prayers so many times that it has become like a routine or automatic practice. These ways of praying will not yield a satisfying result or no result at all because it is the sincerity and the pure and honest intention that matters and what God responds to.

<u>Example #2:</u>

Here let's take the example of a single mom who, overwhelmed by the charges, the bills, and all the problems in her life, one day fell on her knees and started to pray "The Lord's Prayer" and "The Hail Mary Prayer" with so much sincerity, focus, honesty, and truly calling out God from the deepest of her heart and almost in tears asking for Him to give her daily bread. In this second example, the prayer of the single mom will be immediately acted upon by God and answered by the heavenly team.

Can you see the difference between the prayers of these two people? In these two examples both persons are apparently praying the same prayers, but the first person's answer from heaven was, "No. You are not sincerely and truly praying, you are just reciting what you have learned to accomplish your daily Christian routine or duty. We will answer you later when you sincerely ask." On the other hand, the answer to the prayer of the single mom was, "Yes and it shall be," and heaven immediately started to work behind the scenes to help her with her needs and to bring the best solutions to her prayers. Same prayers but different answers from heaven: that is how powerful sincerity and intention are in prayers. For this reason, I will say that when you pray, be really honest with God about what you really want, be really focused when you are praying, and pray sincerely from the depths of your heart with honestly and in humility.

Prayer Essential key #3: the highest good of all involved

Prayers are always answered "for the highest good for all involved". This means that if other people are involved in the situation you are praying about, or your request, Source/Universe will orchestrate the situation and answer the prayer with a solution that will be best for everyone while respecting the free will choices of all involved in the situation. This is because Source/God is benevolent and requests or prayers that will harm someone or that will not be good for someone will be answered with a "No." Notice that I did not say that the prayer will be ignored but that prayer will be answered with a "NO" because all prayers do get answered.

For instance, let's suppose that you are praying that God revenges you and punishes someone who hurt you because you have learned that "vengeance is up to God". So you are praying for the person to die, fall sick, or lose their job; such a prayer is more likely to be answered with a "No" because this prayer is not for "the highest good" of that person, which is involved in your request. I will even add here that this prayer is not even in your own highest good since you don't want to kill or do unloving things because, according to the law of "cause and effect", this unloving wish that you desire for the other person will come back to you and will become a karma that you will learn from later.

Let's take another example: A woman is having issues in her love relationship and she is praying for the relationship with her spouse to get better and to become more loving. Notice that in this prayer, another person, who is the spouse or partner, is involved as well. If this prayer is sincere, it will be answered in a way that benefits everyone and will be "for the highest good of all involved". If it is for the highest good of this lady that she stays in this relationship, guidance will come and the relationship may improve and be more loving. However, if it is not for her highest good that she stays in this relationship, the answer to this prayer may be, "No, there is someone better for you. Get out of this relationship; heal your heart and wait for the Universe to bring you a loving and respectful person that you deserve and desire." In another words, Source is indirectly saying, "Your partner is not a nice person and will not change; get out of this relationship." Now the relationship may get even worse and the sweet lady who is praying may falsely believe that her prayer was not answered or are being ignored, but, in truth, this is not the case. This is why it is important to find a way to tune in and carefully listen to the answers to the prayers as they come because the answer always come one way or another. This example leads to the next Prayer Essential key.

Prayer Essential key #4: importance of "open-ended prayers"

When I was in pharmacy school, we learned about the difference between "open-ended questions" and "closed-ended questions" as well as their importance while interviewing patients. Here is an example of a pharmacist named Todd and a patient named Mrs. Smith in an interview to evaluate how often Mrs. Smith takes a meditation that she was supposed to take every night at bedtime with meals:

Case 1: Closed-ended question
Pharmacist Todd: "Hello, Mrs. Smith, do you often take your medication?"

Mrs. Smith: "Yes, I do."

Case 2: Open-ended question

Pharmacist Todd: "Hello, Mrs. Smith, can you tell me how often you take your medication?"

Mrs. Smith: "Well, I usually take my medication, but sometimes I forget it and remember it only when I am in bed, and when that happens, I just skip it because I am too tired and don't want to get out of the bed and just want to sleep. Other times, when I am not just hungry, I also skip the medication because I am supposed to take it with my supper at night."

In this example, when the pharmacist Todd asks an "open-ended question", he gets the correct answer that he was seeking and that will help him to understand how regularly his patient, Mrs. Smith, is indeed taking her medication. However, when he asks a "closed-ended question" he does not get the correct answer that he was seeking. This is exactly how prayers work. You have to know how to pray and say "open-ended prayers" so that you can have the best answers and the best help from Source/God.

Let's come back to the example of the lady who was asking God to help her in her relationship with her partner. In this case, we can say that she was praying a "closed-ended prayer" because the only solution that she was expecting was the betterment of her current relationship with her partner. However, God can see the bigger picture and is able to see that there is someone else in this world who will be more appropriate for her and that her current relationship with her partner is a dead-end and will not improve because of the obstinate character of her current partner. Thus, God may answer her prayer with, "No, because there is someone better for you." In her case an open-ended prayer can be: "Dear God, please help improve my relationship with my husband or something even better. Not my will, but Your Will, for my highest good and for the highest good of all, May Your Will be done."

Let's take another example of someone who says this prayer: "Dear God, please help me to have money to buy French fries and hamburgers every day." This is another example of a "closed-ended prayer". Instead this person can say, "Dear God, please help me to have money to buy French fries and hamburgers every day or something even better," or he can also pray, "Dear God, please help meet my daily meals and needs with ease and grace." Adding "or something even better" at the end of prayer is a way to state to the Universe your intention that you are open to other solutions, to the best solution. Open-ended prayer is a way of recognizing with humility that you don't know the best answer or solution to your prayer but that you are open and willing for the best solution to come instead of clinging to your own limited view. By adding "or something even better" who knows if God may bring to the person in the last example the finest meals of this world with French vegan food or caviar and champagne from the top chefs of this world and lead him to become a millionaire in order to afford this "something even better" that he has asked for?

These examples illustrate the importance of praying "open-ended prayers" and truly to be open to other ways that your prayers will be answered.

What often happens is that prayers are not generally answered in the way that we expect; this means that the answer to the prayer can come in another form or another way that is different from what we are expecting. However, all prayers do get answered one way or another. Let's say that you are praying to have more money for instance, you may be expecting to win a lottery or to get a raise of salary in your current job, but the answer to this prayer can be different and can come in the form of guidance to take an action such as to start a new creative project or business, to apply for another job, or even to go back to school to do a training that will be aligned with your life purpose and that will yield better, more lasting, and greater results. Guidance to take actions is one of the usual ways that God answers prayers. The key is to pay attention to this guidance, to recognize it, and to act upon it because the outcome will always be good and always yield good results when it comes from Source or God. The guidance is often repetitive and can come in various ways such as intuitions, feelings, signs, thoughts, dreams, conversations from others, etc.

Prayer Essential key # 5: divine timing

Prayers are always answered in "perfect divine timing" and not on human timing. There is timing to everything in the heavenly realm and this is so for prayers as well. Because of the free will law and "for the highest good of all involved" keys, sometimes there can be a delay in receiving the answers to your prayer because God sometimes waits to have the free will "approvals" of others involved and works with them on the spiritual plan to bring about the best solution to your prayers. However, these delays are not truly delays because when the answer comes, it will be in the perfect divine timing. Universe will work with others and the answer to the prayer will come at the appropriate time. What often happens is that when people don't receive the answer to their prayers immediately, they get discouraged and believe that their prayers are not answered. However, the truth is that Universe is working behind the scenes to bring the solution at the right perfect divine time.

Prayers may seem to be delayed, but, in truth, God is simply working behind the scenes, orchestrating things in our favor. In reality, it is simply a matter of time; this means that even though we may think that the answer of a prayer is delayed, the timing is divinely perfect after all.

Prayer essential key # 6: the present moment

Because the only moment that exists in the Universe is the "now" or "the present moment" (the past and the future do not exist in spiritual truth), you must renew your prayer daily or several times a day if needed. Also, you have to pray for everything that you want intervention for.

For instance, let's say that someone just once prays, "God, please help me with everything and for all the problems that will come in my life in the future as long as I shall live. Amen." There is nothing wrong with this type of prayer, but a more effective way will be to pray every day for the things that you need and even during the day as problems or challenges arise. I will say: pray for everything and anything. Pray daily.

Prayer essential key # 7: faith and the importance of directing prayers to the "Divine messengers" that resonate most with you.

Another important element that comes into play when it comes to prayers is trust and faith. The truth is that there are several heavenly beings who respond to prayers. In many cases the guardian angels assigned to you will respond to your prayers and it does matter whether you know that you have guardian angels or if you don't even believe in angels; no matter the religion, the angels are there and answer prayers.

Ascended masters such as Jesus, Moses, the Buddha, Mother Mary, the Saints, etc. also answer people's prayers. God/Source also directly answers our prayers. I will say that it really does not matter to whom you direct your prayers. It does not matter if you call on God directly for protection or if you call Archangel Michael or Jesus or the Buddha. However, since trust and faith are important factors, if you are raised as a Christian, only believe and trust in Jesus, and or do not even know who the Buddha is, it is better to send your requests to Jesus since you trust in Him. Similarly, if you were raised in the Buddhism tradition and do not believe in Jesus, it is better to send your prayer requests directly to the Buddha since you are more connected to Him and trust Him.

The key here is to call upon the Divine messenger or the figure of the Divine that you are more affiliated with, more connected to, and that you believe in. You may also directly call God and the result will be the same. You can call on the Archangels if it is easier for you, it does not really matter.

Furthermore, it is also important to use the name that resonates more with you. What do I mean by this? For instance, if you are more familiar with calling the Divine by the word "God", it is best to use the word "God" in your prayers. However, if the words "Lord", "Source", "Eternal Light", "Universe" are what you are familiar with, it is best to use whatever word works best for you in your prayers. Some people prefer to call Jesus "Joshua"; if this is your case then use "Joshua" when you are referring to Jesus in your prayers. Regardless of your religion, use the word or the name that works the best for you or that you have full faith in when you are praying. Do what works most for you. Your sincerity and desire to get help is what matters the most.

I cover the topic of prayer more in depth in my other book, *"A prayer book for realizing inner peace"*, which expands in great details on how prayers work, who answers them, how to pray, and how prayers are answered and includes numerous prayers for various life situations.

Religion
Subchapter 6

BIBLE TRUTHS REVEALED

What Can You Tell Us about the 10 Commandments?

Moses is another highly evolved being who came to guide, serve, and help humanity at some point in its history. Yes, God did give him what are now called the "10 Commandments" but they were supposed to be suggestions and guidance to follow to assist in life. They were presented originally in this way: "If you commit to these 10 suggestions, your life will be led from the heart and it will make your life on Earth smoother." They were not "commandments" neither are they "authoritative" in any way.

God, the angels, the keepers of the Light, and the heavenly team are not commandants or some sort of authoritative officers or kings who give orders to their subordinates. Keep in mind the "law of Oneness", which means we are all equal no matter our level of evolution, and the "law of free will", which they always respect. This means that we, human beings, are free to do as we please and do receive orders. God, the guardian angels, and the heavenly team only guide, show the way, suggest, and then let humans make their own choices. Furthermore, these loving suggestions have been transformed into something that is now associated with punishment, creating fear in human minds. What was originally "10 suggestions to commit to if you wish to be happy" has become "10 Commandments, and if you don't follow them, you will be punished."

The intent was that if people commit to these recommendations or suggestions, it will ease their lives. God does not command but only suggests, guides, assists, leads, and directs. This is because He gave us the free will choice as a gift. Later, these suggestions or recommendations were misunderstood and now are called "10 Commandments" with a wave of punishment and fear attached to them, while they were simply meant to be a sort of written loving guidance.

God/One Life gave the "10 suggestions or recommendations" to Moses because at that time in the evolution of humanity what was needed was clear guidance, clear and precise suggestions and instructions that people of these ancient times could easily understand and follow. People at that time on Earth were not ready to embrace the idea of "unconditional love and loving your enemies" introduced later by Jesus; that is why this simple, clear guidance was given by heaven. Therefore, instead of telling them, "Love everyone and love your enemy," they were given the suggestion of, "Do not kill anyone, do not steal, do not lie," and so on. It was more digestible for humanity at the time. Keep in mind that even today many of us are still shocked and struggling with the concept of "Love your enemy". However, we are all evolving and getting into better understanding and awareness of love. Today, an increasing number people are becoming more aware and more open to unconditional love more than ever before in the history of humanity.

What Is the Truth about Tithes and Tithing?

Before answering this question, let's honestly do some enquiry ourselves. Does God need or use our money? Where do all of these tithings go? Do the tithings go to heaven? Who uses them? Is it God? Does God need us to give our money to religious institutions to prove our faithfulness to Him? What interest would God have in asking us to reserve or give 10% of our income to organized manmade religious systems?

The truth is that there is a spiritual law that as you give, you will receive. The issue and the untruth is not the concept of tithing itself but the fact that it was distorted and said "to be reserved" or "to be given" only to churches or other institutional religious organizations. The truth is that you are free to give your tithe to people in need or to your favorite nonprofit organizations, your churches or whatever you feel aligned to. It does not have to be to churches necessarily and it does not even have to be 10% of your income necessarily. I believe you can give more than 10% if you wish or less than 10% if you choose to. You can give it to your relative in need, to the homeless on the street, to the poor countries, to sponsor children in need, to help the environment, and so on. The key is giving, helping, serving, and contributing to the betterment of the world. The most important thing is to give with a joyful heart for the love of giving and helping and not through guilt or fear of not respecting some rules of God. There has been so much guilt and so many fears and finger pointing on this topic that have been totally unnecessary and false.

Here is the true law that operates on this earthly plane and in the whole Universe: the law of "cause and effect". Anything you put out will come back to you. When you give, you will receive. When you financially support others in need, you will be supported by the Universe or God or whatever you call the Source that is All-in-All. The money, love, kindness, and compassion that you give to others will come back to you multiplied. This is how it works. Regardless of whom or which organization you give the money to, the most important thing is your loving intention to help and to give to those in need.

The untrue tithing law was devised by groups of humans thousands years ago who instituted that they would be the "only ones" that would collect all the tithes from the population and their congregants. Then these groups of humans stated that they would manage these tithes and money themselves on behalf of God. They created these false ideas and told people that their tithe was for God and that the tithe was not for them but for God. Furthermore, they told people that it was a "law of God" and a "commandment" from God to give 10 percent of their incomes or possessions otherwise the wrath of God would fall on them. The people at the time were afraid and believed these lies since these lies were coming from religious authorities. Thus, the tradition of tithing was created and it continues to operate in humanity today. These false laws were written and somehow were found in some religious scriptures and have been taught to people from generation to generation. These misleading practices and rules became anchored in the human mind and were attributed to God and became a belief that has been practiced for millennia on Earth.

On the subject of tithing, there is nothing wrong with tithing and giving your tithe to churches or other religious organizations. However, the main thing to keep in mind and the thing that matters the most is the intention behind your tithing. If your intention of tithing to your church is to help and to contribute in a way that will benefit others and if the foundation of your tithing is to do something loving for the joy of helping others out of your own free will choice, this is wonderful. On the other hand, if your intention of tithing is to obey "God's law" and follow "His" commandments to implement some sort of points of obedience to God, or if you are tithing out of fear because it is the law of God and you don't want to disobey Him for fear of punishment for not following His commandments, this is another story. Do you see the difference between these two different intentions?

Furthermore, if your intention of tithing to your religious organizations or to any other organization is because you have somehow heard that when you give it, it will come back to you multiplied and you are after this "multiplication effect" and that this the main reason why you are tithing, then your effort is in vain. You cannot fool the Universe or God because the Universe only sees your intention, which is the vibration behind your action or tithing. Giving because you want to receive something in return, the double portion, or quadruple, or 10 times what you have given is very similar to trying to sell something and gain profit out of it while making it sound as if it is intended to help others.

Also if you are giving out of guilt it is also another unfruitful act as well. Therefore, give, but give for the joy of giving, with the intention of helping. Give, but give to the causes that matter to you the most and that make you feel that your giving matters and is making a positive difference in the life of others. Give, but give from your own free will choice not through guilt and not because some preacher has preached about it and you felt bad or because you fear that if you don't give you will disobey some sort of God's laws or God's commandments.

Regarding the question of tithing the answer is simple. Giving is the main key. As you give, you receive. You can give to anyone or any organization that you choose and it is all fine. The most important thing is your pure intention to give. You can choose <u>not</u> to give if you wish to, but as the law of "cause and effect" goes, you will not receive either. Please do not feel guilty if you can't tithe because there is no law, no sin, no condemnation, no consequence if you do not tithe, neither is there any in how to tithe. The door is open. God is 100% unconditional Love, He never judges you or anyone. I would add that God is a God of freedom. Your money is your money and you are free to enjoy, give, and bless others as you wish.

Light of Truth

Several years ago, I had a remarkable discussion with one of my relatives; here, I will name her Angela. During our conversation, Angela told me that one of her good friends who was an engineer lost his job. He was also sick at the time. This couple was having a very difficult time financially. I was really touched by what her friend was going through, but, unfortunately, I didn't have money at the time as I was at school and working part-time as a cashier. However, touched by what this lovely couple was going through, I asked Angela if there was any chance that she could do something to help her friend financially. She replied that she needed to tithe at her church and that, after tithing, she wouldn't have enough money to help her friend and his wife.

I was crushed by that statement. I begged Angela, asking her if she could simply give her tithe or part of it to her friend so that they could pay their bills and meet their medical needs. Angela replied, saying that she was not sure if it was "biblical" to give the tithe to someone other than the church. Again, I was shocked and crushed by her words. Let's us be honest with ourselves; what kind of loving God would order or command people to give money to religious institutions that sometimes do good things with the money such as funding youth and missionary programs but also sometimes use that money to buy new chairs, new audio systems, repaint the building, etc., rather than helping a good friend who is going through health and financial challenges? Does this seem to come from love? Does this so-called "tithing" law seem to come from a wise, all-knowing, and all-loving God? Of course, God did not create any such law. These are all man-made laws and falsely attributed to God and they are untrue.

Who is Mother Mary?

I was very fascinated when I discovered who Mother Mary truly is. Mother Mary is mostly known as the mother of Jesus. However, in truth, Mother Mary is a powerful healer and an ascended master like Jesus who played the role of the mother of Jesus during their world-changing incarnations. She, like Jesus, continues to help humanity to bring healing and love. One of her specialties is helping children. She is a really loving being and often works in conjunction with Archangel Gabriel who also loves and cares for children and their welfare. This is just to say that Mother Mary is not just the mother of Jesus as we may think and that, like Jesus, she cares deeply about the earth, and she is still helping humanity. Mother Mary, like Jesus and the angels, is nondenominational and helps all who ask for her help, regardless of their religions. She has a very loving and comforting energy and brings in a soothing and reassuring peace when you call for her help. She, like Jesus, is often associated with Christianity, but, in truth, they all come for our Source God and are nondenominational as they help everyone equally. Mother Mary is among the main beings playing important roles on Earth currently and who are still with us, helping and guiding this planet.

What Can You Tell Us about the First Man, Adam, and Creation of the Earth?

Adam is mostly known as the first man created by God, but this is not necessarily accurate. I will say that it is true and not totally true at the same time. The earth was created approximately several million years ago or maybe even earlier than that. There were times when there was some destruction by natural disasters, but other humans were reintroduced on Earth again, and these "new humans introduced" can be considered as "the first men" or the "Adams" of these periods. Therefore, there were several "first men" or "first Adams" who were introduced on Earth at diverse periods.

Furthermore, there were souls who were on other third-dimensional planes who did not graduate from their planes and who were brought to Earth to continue their learning and evolution. This is fascinating and explains the diversity of races and ethnicities that we currently have on Earth. When I was young, I used to find it "odd" how God created Adam and Eve and wondered how all the various races were born and started. If Adam and Eve were the only first men created by God there should be only one race currently on Earth: the race of Adam and Eve. However, this is not the case because there are many races on Earth: white, black, Asian, Indian, etc. When I discovered later that various souls from various planets with various races were introduced on Earth in the same way as Adam was introduced but at different times, everything became clear to me. We are indeed in a unique, exciting experiment on Earth, the goal of which is to learn about and embrace diversity and to learn to love others who are different from us.

Did God tell Abraham to Sacrifice His Son?

No. Why would God do that? What benefit would He gain from it? Of course, God did not ask Abraham to sacrifice his son. Keep in mind that God is 100% Love, and, contrary to men, God is egoless and, therefore, anything that does not come from Love or is not of Love does not come from God. It is my understanding that the information has been distorted and changed over time. God has nothing to prove to anyone, and He is not in search of love or approval from anyone. God is God, and He does not even need people to worship Him or do anything to prove to Him their faithfulness: God is all-sufficient, omnipotent, omniscient, and omnipresent. God is All-in-All.

Who Are the Chosen People?

Some spiritual scriptures refer to certain people as "the chosen people", but, in truth, we are all chosen people. No nation is above or below the other, and no person is above or below others. God loves everyone equally, and He does not play favoritism. There were misconceptions and lies that have been circulating that some people are above or below others. This is not the case at all. When you keep in mind that we reincarnate several times and that we can choose different races or genders, it becomes clear that we are all equal. Anything that is not from unity or love is not true. The truth is Oneness, which means that we are all One, and we are all equal. You can see life like an orchestra with various people playing various roles. Yet, like in an orchestra, no player or role is above or below others. It is like a symphony. Everyone is needed, and every part of the orchestra is essential for the symphony. This is the case for life as well. Anything that comes from superiority, inferiority, or being special is not true. Therefore, I will say that there are no chosen people or nation, or, to put it another way: "we are all chosen people."

What Can You Tell Us Regarding the Names or Attributes of God Such As the Ones Mentioned in the Bible?

It is true that the Bible and other religious books mention the names or attributes of God such Elohim, Yahweh, and so on. However, those are not truly the names of God but names of angelic or heavenly beings that God has sent to help humanity throughout history. In fact, Elohim is the name of a group of angels that helped humanity in their evolution in history and that continue to help humanity today. I believe that when these beings communicated with the people in biblical times; they identified themselves as Elohim, which means "the Creator God" and people mistakenly took them for God and said that God's name is Elohim.

Similarly, Yahweh is a name of a group of evolved beings who are in a higher dimension and communicated with people in ancient times; they still communicate with humans today.

Another popular name or attribute of God is Rophe and God is commonly referred to as "Jehovah Rophe", which means "God heals". However, in truth, "Rophe" is simply another pronunciation or version of "Raphael", which means "God heals" or "He who heals". In truth, Raphael is an Archangel, mostly known as Archangel Raphael, who is a "higher rank" angel and who is specialized in healing. In fact, Archangel Raphael is the angel of healing that helped humanity in history and who continues to do so today. Somehow, people of ancient times mistakenly confused all these heavenly beings and helpers for God and there were misunderstandings and distortions that led to what we now know today as the names or attributes of God, which have become popular in some religions.

Can You Explain Some of the Concepts Described in the Book of Revelation?

The Book of Revelation was written in a symbolic, parabolic, and metaphoric way. The truth is that the Book of Revelation is mainly describing the path of the disciple and the challenges that he would have to overcome on his path of awakening to become like Christ. When I say Christ, I do not mean the body/mind complex known as Jesus Christ but the state of Christ Consciousness that we all are meant to get to one day when all our dharma has been completed and when we have become 'enlightened.' As mentioned previously, the earth is not going to be destroyed or to end in any way. There is a bright future for this planet. The earth is going through a dimensional shift from third dimension to fourth or fifth dimension. In truth, there is nothing frightening in the Book of Revelation, as it is simply describing the journey that the hero will take to become like Christ, an awakened and enlightened being; that's all.

To simplify this, below is a table listing some terms and symbols used in the Book of Revelation and what their meanings are:

Terms	What the terms represent and their meaning	Short explanation
The beast	The ego and its lower desires and the negative emotions associated with it that are always in opposition to the Will of God	These are emotions associated with the ego: Fear, jealousy, anger, hurt feelings, regrets, guilt, etc. These are emotions associated with God's Will: love, peace, faith, unity, oneness, service, courage, patience, kindness, etc.

Sign on the arms and hands (or the signs of the beast)	Represents labor and work	Work, labor, career are needed to live and survive on Earth. Unfortunately, oftentimes career and work and the fear of "I am just doing what they asked me to do" or "I am just doing what I need to do to feed my family" are what often prevent the majority of people from following the path of honesty, sincerity, and loving service in life, which is the path that will lead them to awaken to the path of the disciple and to become like Christ. That is why we have immoral salesmen lying to clients or selling things that they know are not beneficial for their clients or lying to make their sales happen. That is why we have immoral pharmaceutical industries or healthcare professionals giving medications that they wouldn't take themselves to their patients. That is why we have so many people trapped in jobs that they hate or doing things that they do not believe in or that they know are wrong. They are all trapped by the fear of 'work and careers.' In a metaphoric way, the writer of the Book of Revelation will say, "They all have the signs of the beast on their arms and hands." Work or labor and concerns about what to eat and how to survive are the first sign (the first sign of the beast) that keeps people captive and blocks them from living the life they were meant to live.
Sign on the head	Represents thoughts and freedom of thinking	The great majority of the population of the earth is controlled by the brainwashing of the media, cultures, rules, and norms that regulate and control people's thoughts and actions, even though many do not even realize or are not aware that they are being controlled and manipulated. Many people live in their ego, the egoic mind, and are mentally in prison. The thought patterns and not being free in the mind and living in the world of the ego in the head and being mentally controlled is the second sign (the sign of the beast) that keeps people captive and slows their awakening to the life of Christ.

Religion
Subchapter 7

THE NOTION OF HELL

What Can You Tell People about the Existence of Hell?

The existence of hell, this is a human idea. There are many who believe in it, and it has become a collective idea. Therefore, we can name this a religion. There is great fear and fear-based ideas surrounding this notion of hell. The true hell is the imprisonment of the mind. Humans are living out the hell here on the earth. Imprisonment of the mind is the true hell.

When you eject or leave the body vessel, all is pure Consciousness; you are merged with pure Consciousness regardless of what you have done in your human experience. All souls are ushered to various degrees closer to merging with Light. There is no such thing as the hell that you go to for bad behavior; this is a human idea. Release yourself from the torture of thinking that you will be punished. Spirit or Light or God does not know punishment or hell; simply "cause and effect." The karma or dharma of your existence will be played out in accordance with your awareness or understanding. This is why we return to Earth, incarnation after incarnation, again and again, to learn, not to be punished but to learn. "Cause and effect" is quite separate from going to hell.

The fear-based idea of hell; please, do not say this to children. Please, do not instill this false program into a child's mind because it creates great fear, which into adulthood is the cause of disease. Instead, teach children that there is, in science, the law of 'cause and effect.' What you put out will return back to you, and therefore you are mindful of yourself. Your actions and your thoughts and feelings and what you are giving out to others is what you will receive, and life is harmonized and balanced that way.

When you leave the body (meaning when you die), everybody is invited to the pure Light and Love; Consciousness is Light. Every soul will have a choice, as with free will, how close they wish to merge with Light or if they would like to hang back. For instance, let's take the example of Adolph Hitler; there is a collective idea on our planet Earth about this former human being. Even Adolph Hitler, who is now in pure Consciousness and has continued to evolve, after his death, was provided the opportunity to merge with Light. There was no punishment for him at all. However, the soul did decide to hold itself back and not to merge itself completely with pure Light. Why? Because it needed to feel itself learning more and return again and again.

The karmic wheel will spin, and it will continue to do that lifetime after lifetime until the soul has completely merged, merged to the top tier of Light, and when this occurs, the soul has the opportunity to serve as a guide, Saint, or guardian angel and can remain in pure consciousness without reincarnating if it so chooses. This is the pattern process for every human being, the process of evolution, an incarnation that every soul goes through regardless of the crimes committed in the human experience.

Where Is Adolph Hitler Now?

He is in pure Consciousness, pure Light and Love. This is how benevolent Source/God is.

Light of Truth

We have been told so many lies, especially regarding the existence of hell, but somehow we believe them. We have been told that God is unconditionally loving, and yet we still believe that He will throw us in eternal fire if we behave badly on Earth. However, even the worst parents will not throw their children in a fire and watch them burn if they behave badly. Do they? If you are a parent, and you are reading this book, could you throw one of your children into a fire? How awful is this idea? Even the worst criminals among us dearly love their children and want good things for them. How, then, can we as a society buy into the lie of the idea that we will be punished and thrown into hell? Alas, the existence of hell is one of the biggest lies of all time that we have been told and that continues to create great fear today.

The example of Adolph Hitler may be very shocking for many because, somehow, in our human thinking, we think that Hitler deserves some sort of punishment. However, the truth is that there is no punishment for anyone, including Adolph Hitler. The same law of Love applies to everyone, and the same law of "cause and effect" applies to everyone regardless of what we have done or not done on Earth.

I would like to add here that the real tragedy regarding the case of Adolph Hitler is how far the madness went before the community of Earth started to react. The real tragedy regarding the genocide of Rwanda is how many people have died and how far the madness went before the international community intervened to help stop the atrocities. The real tragedy regarding people dying of famine is to watch the whole world remain silent because we, as a community on Earth, are unable to come up with a plan to stop the madness of watching our brothers, sisters, and children dying on the other side of the sea from lack of food. These are the real tragedies.

The Truth about Life

CHAPTER 10

RELATIONSHIPS, SEXUALITY, MARRIAGE, DIVORCE, UNFAITHFULNESS, CELIBACY, ABUSE

"A dog barks when his master is attacked. I would be a coward if I saw that God's truth is attacked and yet would remain silent."

–John Calvin

Sexuality, Marriage, Divorce, Remarriage, Celibacy

What Can You Tell Us about Sexuality?

The topic of sexuality is an important topic, but there is so much confusion, misunderstanding, and misinformation associated with it. For millennia, sexuality has been associated with taboos, shame, guilt, and even sin. We have been told that sexuality was created by God for the purpose of procreation only and that it is immoral or sinful if used for other intent than procreation. However, this is not true. God has not commanded anyone to experience sexuality only for procreation or in the parameters of marriage only, and all these false ideas, taboos, restrictions, shame, guilt, and blame associated with sexuality were all created by organized religions and cultural dogmas with the intent of creating guidelines for propriety and morality or simply for the purpose of control millennia ago. The truth is that sexual energy union is a wonderful thing, a gift from God that is meant to be experienced and enjoyed freely, not just for the purpose of procreation or only just within marital parameters. Anything that God has created is perfect and created out of love, and God has not created anything that is shameful or immoral. Sexuality is meant to be a celebration of love. When experienced between two consenting people in this way as a celebration of love, how can that be a sin, shameful, or immoral?

Today, between the cultural rules and lies around the topic of sexuality, the false teachings of the churches (which state that sexuality was created by God for the sole purpose of procreation), the shocking use of sexuality for exchange of money such as in prostitution, the gain in popularity of pornography on the internet, and the addiction to pornography that many are caught in, the confusion around the topic of sexuality is enormous.

The intention and the purpose behind anything is very important, and it is the thing that matters the most. When practiced with kindness, love, compassion, respect, and between two consenting people who are in love with each other and who deeply care for each other, sexuality is indeed a wonderful gift from God, and it is meant to bring to these two people the experience of oneness, love, and help them to get closer to one another. Therefore, it is not sinful or immoral for two people who are in love to have a sexual relationship outside the boundaries of marriage. There is no spiritual law against love. How can love or sexuality be a bad thing or shameful or immoral? The sexual energy is a divine energy. It is an expression of love and the experience of divine love while in the physical body. There is no sin, shame, or taboo attached to it. All other things such as the rules, shame, blame, fear of punishment associated with sexuality are all created by humans themselves, and these ideas and misinformation have led to so much pain, confusion, suffering, and guilt in this world. These false ideas and beliefs have detoured sexuality from its original purpose of joy, celebration of love, remembrance of Oneness, and divine love.

What Can You Tell Us about Marriage, Divorce, and Remarriage?
The answer to this question is simple and can be summarized is one word: 'freedom,' 'free will.' We are free beings coming from the Divine, and we are free to choose whatever we desire for our love lives: marriage, celibacy, divorce, or whatever. There is so much misinformation on these topics. Two people are free to make agreements and consensus between themselves, which they refer to as marriage, dating, engagements, and so on. At the same time, if one of the parties involved does not like the terms of the agreements anymore or has simply changed his or her mind, then they are free to break the agreement and set themselves free. We are free beings. Keep in mind that some souls even decide to incarnate and then simply change their minds and return home to heaven, and their choices are respected and honored. Why, then, will God or anyone go against their free will if they have decided to get out of a relationship that they do not like anymore? Why should it be sinful or unlawful to break an agreement of what we called marriage if the agreement is no longer pleasing or suitable for the persons involved? There is no sin in leaving a relationship that we do not like, and God did not make these laws. These laws, regulations, and expectations regarding marriage are solely man-made ideas and laws that have been attributed to God and recorded in some spiritual scriptures as 'God's laws,' 'God's expectation,' or 'God's

commandments.' Regarding remarriage, under the law of free will and freedom, we are free to get remarried, to be single, or to live freely with love partners without any fancy certificate or official marriage. We are free.

The other thing to keep in mind is that we are constantly changing and evolving, and sometimes we can outgrow a relationship. This means that we have changed, but the other person did not, which is fine because everybody is on a unique path with no comparison. Therefore, since we are in constant change and evolution, it is normal that relationships will start and end. When you change, it is more likely that your environment and your relationships will change as well. For instance, let's suppose that John and Mary met each other at high school and were so in love that they have decided to get married after their graduation in high school. When they were in college, they had the same interests; they liked to eat outside and taste various delicious barbecues and drink beer. Several years later, Mary has changed, become a vegan, and has no interest in traveling to taste meat-laden foods. She has decided not to drink beer anymore and prefers spending quiet time meditating and reading inspirational books instead of fleeting pleasure and beer. Suddenly, John and Mary started growing apart, and they started to have many arguments about several things in diverse areas of life, and their marriage became tense. Indeed, they started to have several issues that are tearing them apart. Would it be sinful, bad, shocking, or unlawful if Mary decided to break her agreement and ask for a divorce? No, there is no sin in that. It is simply part of life. Physical forms, including money, material objects, jobs, and people will come and go in life. This is the dance of life. Life is made of ups and downs. Things will come and go. Marriage will form and will end. Babies will be born, and people will die. This is just how life is: an unpredictable ride when you never know where the wind will blow and what will come tomorrow.

Also, sometimes some relationships simply end because their purposes have been served. Sometimes we have lessons (karmas) that we have elected to learn, and when the purpose of the lesson has been served, the relationship ends.

In our western society, and probably in other civilizations, divorce is feared so much that people cling to unhealthy marriages or even get depressed at the simple idea of being divorced and living alone. However, life is always about positive changes and always goes cyclically. Just as the weather has various seasons such as winter, fall, spring, and summer, there are also various seasons in life as well. Sometimes, seasons of life end simply because it is time for a new chapter, a new rebirth, and the 'old' must go in order to make room for the 'new' to come. The winter must go in order to make a place for the spring to come.

What Are Your Thoughts about Being Single or Getting Divorced and Being Single Again?

Nobody likes to be single, or at least very few people do. No one wants to sleep alone in a cold bed at night, coming back from work and having no one else to talk to, eating alone, and going through life alone. However, we are never truly alone because we have our guides and angels who are constantly around us even when we cannot see them. Nevertheless, the feeling of being alone seems real and heavily weighs on us. When you see people snuggling with their lovers, or their loved ones, or their children, it is impossible not to want the same thing or to feel alone and sometimes even have the feeling that we are left behind in this life. However, being alone, even if it is temporary, and taking good care of yourself and your heart, and taking inventory of life, is sometimes better than jumping into the first relationship that presents itself to you just because of the fear of being alone.

Oftentimes many people who are single feel like they are victims of life or that they are unlucky or simply can't seem to find 'the good ones.' Furthermore, those who were married and going through divorce are even more afraid of being alone. Being single has become like a plague that we want to avoid at all cost. This is because we are afraid of being alone with ourselves, and we have developed beliefs that we need someone or something to complete us. However, we don't need anyone or anything to complete us because we are all created as whole and complete. It is possible to be at peace and content and fill ourselves with so much self-love that we are at ease, at peace, and content with ourselves even if we are single. In fact, it is this state of being that we must reach before even contemplating the idea of entering into a new relationship so that we can enter the relationship with fullness, filled with our own self-love, instead of entering the relationship in neediness and with hope to find someone who will complete us or fill some void that we may have.

The truth is that we, human beings, are creatures of habit, and when faced with change such as the idea of getting divorced or being alone, we panic and automatically resist. We are comfortable with what we are familiar with and afraid of the unknown, and when faced with divorce, we wonder what will happen and if we can make it without our partners. However, change is how we evolve and grow in life. There is nothing scary or dramatic about being single or divorced. If you have been single before (which everyone has), then being single again is not that different. Change does not have to be scary, and a divorce is not the end of the world. Only the ego (the thoughts in the mind, the barking dog) makes divorce seem bigger and more difficult than it actually is. Divorce is seen as a failure: "You have failed in your marriage" or "You have failed in your love life," the ego or other people will tell you. This misperception in our society places a huge burden on those going through a divorce.

THE TRUTH ABOUT LIFE

What are my friends, coworkers, or relatives going to think or say about me if they learn that I am going to get divorced? we often wonder. However, divorce is nothing else but a change, an end of a season, and does not have to be associated with guilt, failure, shame, or blame. It is just a change, and it can be a positive change.

What many of us are struggling with is self-love, and we believe that someone has to give us love or that we need someone to love us. However, only you can love truly and completely love yourself, and I will say that unless you learn to love unconditionally for who you are right now, you will run all your life searching for someone to give you that love, and you will never find it because true love begins with the love of self. Your partner cannot give fill that void that you have inside of you, and only you can fill your own tank with love first. Searching for someone to love you is not true love; it is codependency and neediness. Codependency comes from the desire to fill a void inside and the search for an outer source of affection and attention as a solution. No one can ever truly fulfill you and make you feel loved. You and only you have the power to give to yourself the love that you are seeking.

The truth is that what we all are truly yearning for is the complete feeling of unconditional love and safety, the absolute sensation of being loved for who we are, the feeling of total acceptance, and the complete merging with divine love. Therefore, regularly tuning in and connecting with God in calm meditations or prayers can help fill our love tank instead of looking for outside sources such as our partners, children, or food to fill us up, which they can never do. Otherwise, we will always be asking questions such as "Where can I find someone who will love me?" or if we are already in a relationship, we may be asking questions such as "Do you really love me?" or getting upset because we say "I love you" to our partners and children, and they do not say "I love you too" back to us.

Be your own valentine and start to learn to love yourself for who you are right now. Every day, you can tell yourself, "I love you. What can I do for you today to make you happy?" Pamper yourself; buy yourself flowers and gifts; treat yourself as you would an honored guest; do what makes you happy and start to love yourself for who you are right now. Keep in mind that before being with someone, you must learn to be alone with yourself and to be happy with your own self first. Love is not outside of you, needing to be sought after, because you are the embodiment of love itself. Furthermore, as you learn to love yourself, you will easily attract more loving commitment from others because attractiveness shines brightly from those who love and appreciate themselves.

What Can You Tell Us about People Who Do Not Have Children?

Not having children can be challenging for some, but it is not the end of the world. I believe that throughout our many incarnations, everyone will have to experience various things, including being a mother, a father, a sister, or being without children. Also, keep in mind that children are souls who are in our care for their times of incarnation and that love can be shared in various ways. When we truly understand this principle, we will realize that the only thing that matters is the love shared between the child and the parents, and it does not matter if it is a biological child, an adopted child, or a sponsored child. Also, it is important to remember that some children, before their incarnations, chose that the parents who were going to raise them would be their adoptive parents, and yet they choose to incarnate with other biological parents before ending up with their adoptive parents, and their adoptive parents have agreed to this as well.

Also, sometimes souls choose not to have children during specific incarnations because they have decided to dedicate themselves to be in service to others. This is a noble decision indeed, but, of course, once in human bodies, they will not remember that they have elected that for their incarnation, and some struggle with the fact of not having children, and others try desperately to have children on their own. This can be one explanation. It could also simply be that the soul has decided to learn how it is with not having children or for several other reasons.

Light of Truth

I was surprised the first time I learned the spiritual truth that some souls decide not to have children and to be in service to others, but I remember that one of my aunts did not have children, and she struggled her entire life with the fact of being the only one among her sisters and brothers who does not have children in the family. She always wanted to have children. However, we, her nephews and nieces, were like her children, and she supported us financially and emotionally with our education, and she helped a lot of other children and families in the neighborhood and in her workplace as well. I always felt like she was a pillar that supported and helped the whole family. Decades later, when I learned that sometimes souls choose not to have children during their incarnations because they elected to serve others, the story of my aunt came to my mind, and I remembered how she dedicated her life to helping us and supporting many people.

What Can You Tell Us about Celibacy or the Vow of Sacrifice for the Sake of Serving God or the Church?

We do not need to sacrifice anything. We do not need to abstain from marriage or sexual relationships in order to please God, to serve God, to serve the church, or to serve anyone or anything. We do not need to sacrifice our happiness, freedom, body, pleasures, or dreams in any way. We do not have to sacrifice anything. In truth, what God desires for us, for all of us, is to be happy in all ways and to live up to our potentialities. Heaven does not want anyone to sacrifice their happiness in order to help others.

The ideas of sacrifices or making vows of celibacy to God or Jesus or to whoever were created by organized religions and are untrue and do not come from God but from humans themselves. This misinformation and untruth has led to so much confusion, suffering, and shocking stories of priests having mistresses or affairs with other women or committing abominable acts. The basic scheme embedded in this misinformation is the idea that someone who has pledged a vow of sexual abstinence is purer than someone who is in a relationship, thus pointing to the fact that sexual relationships may be somehow impure, dirty, or sinful. However, the truth is that making these vows does not take away the fact that these priests, nuns, or sisters are still sexual beings who have sexual desires. Why not let them get married so that we can see an end to these self-deceptive vows and games? These lies must come to an end. God does not require anyone to sacrifice anything as proof of fidelity or devotion to Him. These are pure lies started millennia ago by organized religions, which still remain today, destroying the lives of young people who engaged with the church in the false hope of serving God or serving humans. God did not ask anyone to vow a pledge of celibacy to Him in order to serve. Can't they still serve God while being married or having loving relationships with their partners? Why do they have to vow a pledge of celibacy in order to serve God or to serve people? This is something that all humanity has to reflect on.

Infidelity, Polygamy, Abuse

What Can You Tell Us about Polygamy or People Who Are Unfaithful in Relationships?

Oftentimes, people carry the memory of past lives with them in their subconscious and memories of ancient times and civilizations with them. Some may have been in polygamous relationships or have had many partners in previous lifetimes in civilizations where polygamy is acceptable and the norms of the society. Therefore, when they incarnate in these modern civilizations, they carry these memories with them, and they have trouble being in monogamous relationships and sticking to only one woman. This may happen also if they have incarnated many times in parts of the world such as the Middle East or Africa when polygamy is normal, and they have had many spouses and enjoyed the pleasure of having multiple partners. Therefore, they have the desire to be with many women because, somehow, in their subconscious mind, there is nothing wrong with that; they just wanted to be with many women at the same time as they used to be in their many previous incarnations. However, if they have now incarnated in this current lifetime in societies such as ours in the United States where monogamy is the norm and polygamy is condemned, they often try to convince themselves that they are monogamists, but at the same time, they may have affairs or extra-conjugal relationships, which they may try to hide, since it is considered as cheating and shocking. They may be also inclined to have several partners. This concept and revelation may be surprising and shocking for many in our western society, but it is just the truth.

This paragraph is not to justify any unfaithful behavior or polygamy or anything of that nature; it's simply to bring light and clarification to some love relationship situations that we often do not understand.

THE TRUTH ABOUT LIFE

People who are in relationships and are dealing with these issues of unfaithfulness have to make the decision of whether or not it is a deal breaker for them. Of course, everyone has the free will choice to decide whether or not they want to stay in a relationship, what their expectations are, and what is acceptable or not for them. If cheating is a deal breaker for you, and you are unhappy with the relationship, then leave the relationship. Someone more appropriate for you and who understands the principles of faithfulness may show up; you can manifest another relationship.

There is no right or wrong relationship type in spiritual truth or in the eyes of God or the angels. Whether it is monogamy, polygamy, or other types of relationships, they are simply what they are. They are simply choices. There is no right or wrong choice; there are just choices.

Polygamy partnerships present more learning opportunities. When I say 'learning opportunities,' I mean opportunities to learn to love the other parties and have more life lessons embedded in them. Think about this for a second: How difficult or challenging would it be to love the other wives of your spouse in a polygamous partnership? Loving and accepting the other wives of your husband or learning to love the woman with whom your husband has cheated, I mean, it is greatly challenging for many, and it takes great spiritual understanding and maturity to come to that point of love, understanding, and acceptance. Yet, some have chosen these particular scenarios for learning purposes and for their spiritual growth. No one has chosen that for them but themselves. Again, this is not a condoning or condemnation of any type of love relationship system but simply the spiritual. This is not also a means or a pass to accept any relationship or situation that someone does not like. Remember that everyone has the free will choice to walk away from any relationship or situation that does not resonate with them or bring them happiness. The choice is ours at any given moment, and we have the power and freedom to change our minds at any time and to choose something else during the course of our incarnation.

What Can You Tell Us About Abusive Relationships?

Abuse of any kind is not acceptable. For people who are in abusive relationships or marriages, it is most likely that their partners have been their parents or caretakers in past lifetimes and that they had karmas (life lessons) to work on and to resolve. Also, people in abusive relationships are probably learning how not to be abused, how to set boundaries, and how to find a way to stand up for themselves and get out of the abusive situations.

I want to reiterate here that no one should accept abuse of any kind, and there is no excuse for abuse, and if you are in an abusive situation, please contact the law enforcement authorities, call for help, tell others, and do whatever you can to get out of the situation or the relationship. It is not God's will for anyone to suffer, and God only wills peace and happiness for everyone.

Sometimes what keeps people in these relationships are financial concerns, the fear of the unknown, and so on. However, the truth is that God has given each and every one of us gifts that we can use to support ourselves financially during our earthly lives and for the rest of our lives. No one is without a gift, so please find out what gifts and talents our Father-Mother-God has given you so that you may use them to free yourself and be financially free. Every single person has skills and talents that they can use to support themselves financially and to be independent. There are simple ways to discover what your talents, gifts, life purpose are in life. Please, if someone is in an abusive relationship, do not be ashamed or afraid to speak up. Ask for help; call the law enforcement in your areas; tell your relatives; do your best to leave the relationship. Universe will find a way for you, and all will be well at the end.

I may be writing a small booklet of simple exercises to discover your life purpose later; if you are interested in it, please look for it in the future.

Self-love

How Do We Love Ourselves?

First, before we can love ourselves, we have to know the difference between our true self (the soul) and the false self or the ego self, which is nothing but a projection, an illusion. Many of us do not know the principal difference between the true self and the ego self, and we have come to believe that we are what the ego tells us that we are, such as the loser, the smart one, the engineer, the cashier, the unworthy one, the white, the black, the spouse, the son, the wrinkled, or the fat person that we see in the mirror, which are all false projections and labeling of the ego.

The true self, the true you, is that which is indestructible; the soul, the Essence, the Energy, the Presence which is beyond the labels, beyond the body, beyond age, beyond career, beyond skin color, beyond accomplishments, beyond your name. This is who you truly are. The only thing that you can say about yourself that is true is "I am." Any other thing that comes after "I am" is a label that changes with time, with circumstances, with training, with education, with location and really changes from lifetime to lifetime. For instance, you may say, "I am 35 years old"; yes, this may be true for now, but the next year, you will no longer be that age. You may say, "I am a bank teller," and it may be true for today, but tomorrow, you may be fired and find yourself without a job and working in another job months later. You may say, "I am a mother," and it may be true, but God forbid tomorrow your children may die in a car accident, and suddenly you are no longer a mother. You may say, "I am poor," but who knows, tomorrow you may win the jackpot of the lottery and become suddenly rich; it is probable, rare but probable. You may say, "I am a black woman," but this is just a description of your physical body, the vehicle, and who knows, in your next lifetime, you may be a white man somewhere. Therefore, fundamentally, the only thing that you can say about yourself that is true always is "I am."

The difference between the true self and the false self (the ego self) is very clear, simple, easy, and pure like water. The true self is the soul self, which is indestructible, unchangeable, the energy, the 'I am' that prevails beyond time, space, labels, skin color, and anything else. For more information or clarification regarding the false self or the ego self, please see the chapter about the ego.

Once we have identified the true self, the question is, how do we love ourselves? Our parents did not do this very well for us, and most human beings can say this, so it is time to become our own parents and love ourselves. It is not selfishness. Not in a self-centered ego-driven way, but rather, to love one and to care for oneself is to provide a parenting, a motherly and a fatherly love to ourselves every day.

Self-love is the pillar of many other things in life. For instance, before you can manifest abundance and wealth in life, you must love yourself to some extent and believe that you are worthy of great things and that you deserve greatness and good things in life. It is this self-love, self-value that becomes the fuel that pushes us to even dream and try for greatness in life. Those who lack self-love can't even imagine doing something other than their current jobs and can't even imagine money flowing to them in large quantities all the time doing what they love. You see, self-love is linked to deservingness and worthiness. Fortunately, it is something that can be cultivated, and everyone can change and is able to cultivate self-love.

Also, self-love is the basis of the type of relationships that you attract and allow in your life. Those who lack self-love will be likely to stay in unloving or even abusive relationships because they unconsciously think that it is the best that they can have or haven't seen anything better in their lives before or simply for fear of being abandoned. Therefore, they often think that they have to tolerate the unkindness, mistreatment, or abuse. Indeed, the lack of self-love leads many of us to put up with bad behavior, mistreatment, and even abuse in relationships. This is another reason why we must cultivate and learn to love ourselves. No one else can do this for us, and it is never too late to start loving ourselves.

Below are some suggestions for how to love yourself

- Take yourself and hold yourself. Give yourself a squeeze and tell yourself, "I love you."
- Surround yourself with supportive and loving people and allow them to give you a hug and a kiss, and you give them hugs and kisses. In fact, there are many scientific studies that support the healing effect of hugs.
- Calm yourself with meditations every day.
- Feed yourself with joyful, bright, and happy food. By bright and joyful food, I mean fresh produce and natural foods.
- Help yourself by giving yourself clothing and shelter.
- Protect yourself. By loving and honoring yourself, you protect

yourself and your energy.

- Listen to loving affirmation statements every day. Listen to the sound of love, loving affirmations, and the sound of the ocean. Let the ears digest love.
- Be in contact with plants or animals or children every day.
- Release all self-blame, self-judgment, and self-harm.

Other suggestions for self-love

- Pursue your goals and interests independently of what others think or believe, even if no one supports you.
- Forgive yourself with compassion, knowing that at each moment, you are doing the best that you can with what you know. Forgive yourself for everything; I mean, everything.
- Talk kindly to yourself in your mind. Be gentle to yourself in your words and actions in all ways.
- Let go of all self-punishment tendencies, behaviors, and negative self-talk.
- Accept yourself with approval and compassion, no matter what you think you have done in life or have not done.
- Once in a while, do something nice for yourself. Do what makes you smile.
- Pursue a creative passion, whether it is writing, cooking, music, art, or whatever it is. Find out what creative outlet is best suited for you and do something every day in that direction, even if it is just for 15 minutes.
- Go outside daily and take a walk, even if it's just for 20 minutes. It will clear your mind and relax you.

- Surround yourself with loving people. Be selective. Know and affirm that you deserve loving relationships and pray for assistance if needed.

CHAPTER 11

GENDER, TRANSGENDER, PERSONALITY TRAITS, FEELINGS, AND EMOTIONS

"I prefer to be true to myself, even at the hazard of incurring the ridicule of others, rather than to be false, and to incur my own abhorrence."

–Frederick Douglass

Gender, Transgender

What Can You Say about Gender?

The truth about gender is simple. We all have both male and female energies inside of us. In other words, we are not solely male or female; we are both male and female at the same time. During the incarnation, we elect or choose a body vehicle that is either male or female and that is designed to best serve our life purposes. However, whether we have a male body or a female body, in reality, we have both strands of male and female energies inside. For instance, a soul whose life purpose requires a great deal of intuition and nurturing or who wants to be a mother may choose to incarnate as a woman, and a soul whose life purpose will require determination or a great deal of physical strength may elect to incarnate as a male.

Both male and female energies need to be in balance for a fulfilling life. The macho man is not expressing the nurturing and feminine part of himself and a weepy woman is not expressing the strength and assertiveness part of herself. In fact, when both male and female sides are in balance, we are more in tune with ourselves and our souls. The female side of us is related to intuition and nurturing, while the male energy is related to taking action, taking control, and so on. The goal is to express both qualities in a balanced way. Our souls have no gender or, to be more precise, have both genders. Every person has both male and female energies inside of them. Some express one side more than the others, and the key is to get to a place of balance.

For millennia, the female gender has been negatively portrayed as weak, inferior, and so on. These are false beliefs and propaganda that come from ignorance and misunderstanding of the value and the uniqueness of each gender. Fortunately, things are changing for us on Earth, and we are starting to realize the equality and value of each gender. Today, men are being encouraged to be open and to express their feelings and nurturing sides, and more women are taking leadership positions and other roles that once were thought to be reserved for men only. We can acknowledge that we have made significant progress on that matter and that many people now understand that women are not inferior to men, or weak, and have started to embrace equality. However, we have more work to do because there are still many injustices or limited beliefs regarding women on this planet, which all come from misunderstanding and disconnection from the truth.

Can You Give Some Thoughts and Insight Regarding the LGBT (lesbian, gay, bisexual, transgender) Gender Identity Phenomenon?

What happens sometimes is that some souls may have 2,000 incarnations in a specific gender, and when they incarnate in a different gender for the first time, they may struggle with their gender identity because it is unfamiliar to them, and they may have challenging times adjusting to their bodies and their new gender. Let's suppose that someone has incarnated 1,700 times consecutively as a woman. However, this lifetime, he decided to incarnate as a male because that is what will be best for his life purpose in this lifetime. Now, this person is incarnated as male but feels really uncomfortable in his body because, somehow, he wanted to be a woman and is attracted to dresses, makeup, long hair, and even acts as a woman. When he was young, he said things such as "I do not feel like I am a boy" or "I feel like a girl," and he likes to play with baby dolls, likes ballerinas, the color pink, and so on. The reason is that this soul may be having trouble adapting to his new body and may be confused and have memories of being a girl or a woman.

The LGBT (lesbian, gay, bisexual, transgender) gender identity phenomenon may be the case of people who feel like they do not belong to their gender or that they are of a different gender. I hope this brings some clarity to the topic of gender.

It may also be possible that some may elect these experiences themselves just to know what it is like to have different sexual expressions in society.

I would like to mention here that it is probable that we all have experienced various type of sexuality during our many incarnations. Sometimes society may approve of our sexuality, and sometimes it does not. They key is to be at peace and accept ourselves as we are. I also would add here to be compassionate toward everyone. We do not know people's stories, their backgrounds or past life experiences, so we should not judge anyone. Why discriminate or reject them just because they have a different sexual orientation? There are places on this planet where they are even killed because of their sexual orientation. What wrong have they done to others? Why judge them and treat them poorly? For those who are angry about them or judge them, what wrong have they done to you? You live your life, and they live their lives. This is the same destructive belief that says some races, the blacks, the Mexicans, the Arabs, etc., are less than or inferior and need to be discriminated against or poorly treated. This madness must stop. Let them alone. Love, compassion, kindness, and acceptance are what we need in our world. We do not have to agree with someone but just accept them as who they truly are: they are children of God experiencing various types of sexuality during specific incarnations; that's all. Let people be who they are and let them live the experiences that their souls have chosen. Let's choose peace and acceptance for all.

Personality Traits, Feelings, Emotions

What Is the Truth about Personality Traits?

Personality traits are what make us unique and different at the same time. They are also what create the most challenges in relationships. When we have issues with others, or if you hear someone criticizing others, what they are criticizing often is the personality traits and the external makeup of the person, not the real true self or the divine self of the person. The divine self or the true self has no gender or personality traits, even though it has its uniqueness. Focusing on people's personality traits is focusing on their egos to some extent, and it only brings division and strains in relationships. As stated earlier in this manuscript, we are made of light, the light of God, and there is a divine light within everyone, no matter their exterior appearance.

There is indeed a way to practice and see the divine light within people instead of seeing the body or focusing on their external appearance and personality traits. It comes with practice, and the more you practice it, the easier it becomes, and the smoother your relationships with people will get. How easy it would be to get along with others if we all had the same personality traits and characteristics. At the same time, how boring life would be if it was so! Personality traits, like the body itself, are not who we truly are but parts of the master design illusion that enables us to have human experiences and to see ourselves as unique and separate from others.

The fundamental truth is that we are not the body or the persona that we think we are, and this shows how illusory life is, and it helps to see life from another angle, another perspective. Life can be compared to a playground, with little children with different personas playing various roles for a little while and then returning home to heaven with the lessons and the experiences that they have gathered during their playtime. This is the gift of God. To come, play, and at the end, you go back with a ton of learning experiences, lessons, and growth.

What Can You Tell Us about Feelings and Emotions?

Feelings and emotions are parts of the human experience. They are not who we are; they simply come and go like many things in life. However, they are so compelling, and we often believe them or act upon them. Nevertheless, the only trustworthy guidance that we have is our intuition.

There are ways to deal with negative feelings and to heal them. Staying with the feelings and simply acknowledging them helps to soothe them. However, when we repress them or try to run away from them, they will bubble up in another form later. We do not have to apologize for our feelings; simply acknowledge them and stay with the feeling. Healing our feelings and emotions takes some willingness and work, but it is the best way to deal with them. Oftentimes, feelings are fueled by the thoughts about them or the fact that we do not want to have these feelings. The more thoughts and attention we give to the feelings, the more compelling they are, and the more challenging it will become to address and heal them. The truth of who we are is beyond feelings, emotions, thoughts, personality traits, and so on. Feelings, emotions, thoughts, personality traits, and even our bodies are like machinery and complex systems that come with the fact of living in human form. Beyond these seeming differences and aspects that make the human experience challenging and unique at the same time, we are pure Consciousness, pure Essence of God. Knowing that feelings and emotions are not who you truly are and that they are simply like winds that blow and cease helps to put them into perspective.

The Truth about Life
CHAPTER 12

CHILDREN

"The pursuit of truth and beauty is a sphere of activity in which we are permitted to remain children all our lives."

–Albert Einstein

What Is the Truth about Children?

As mentioned before, we reincarnate many times. With this in mind, children are not just children. The soul of a child may even be 'older' than his or her parents' souls, even though souls are ageless in truth. For instance, a child may have 2,000 incarnations, while his or her parents may have less than 100 incarnations. You can sometimes recognize these 'old soul' children by the way they behave in life, as their wisdom shines through. Sometimes people even say things like, "This child seems very mature for his age!" or "This child is so compassionate and deeply cares about others," and they are right! What they are feeling is the ancient wisdom and maturity of these children's souls. However, whether a child is an old soul or a young soul, it does not matter; we all come here several times, and children are not just children; it is better to see them as souls and treat them with love, respect, and affection.

Nowadays, children are coming more and more prepared and equipped for their incarnations and diligently choose their parents and caregivers. Once we are born, we are conditioned by society to follow societal rules or to think or behave in certain ways instead of being our authentic selves. Therefore, children are coming more prepared to resist these conditionings and societal rules so that they can follow the desires of their hearts and accomplish their life purposes. That is why many young children are eager to say "No" to their parents in their early ages. They do not hesitate to say "NO" or to resist rules or following directives because they were well prepared before their incarnations to be their authentic selves and to enjoy their freedoms. Also, children are more in touch with who they truly are and with their life purposes, as they have not been deeply conditioned by society yet. Unfortunately, children become molded and shaped by societal norms and dogmas as they grow up and end up forgetting who they are and why they are here.

Children also have greater awareness and remember better their connection to heaven and to the Divine. They sometimes are able to feel or even see the presence of their angels around them. Their sensitivities, simplicity, sincerity, and openness enable them to be more in touch with their divine nature than adults. However, as they grow, they become conditioned by society and their environments and end up forgetting their divine nature and their natural marvel of life of their childhood. There are many stories of children reporting seeing an angel or saying things that surprise their parents. Oftentimes these reports are true because children are more in touch with their divine connection and may still have their third eye, which is the true eye of the soul open, which enables them to see their angels and spirit guides. It is best not to consider children as 'little humans' or 'less than adults' because, in truth, they are nothing less than a grown-up adult and may even be more advanced and mature spiritually than their parents or many adults. The best way to see children is to consider them as sacred divine beings and as important as any other adult.

The sad thing is to see innocent young children being abused by adults simply because they are children and because people wrongly think that children are weak and defenseless. Therefore, many careless adults use their power to control and abuse these young souls. These erroneous beliefs have led to destruction and traumatic childhood experiences for many children, which they carry through their adult years. Even being harsh verbally, screaming at children, using force to make them comply, being too strict with them, or not hearing their voices and honoring them have profound negative detrimental consequences on children and affect them deeply because they are sensitive, and their souls are pure.

In a recent statistic of 2019, the American Society for Positive Care for Children (American SPCC) reported that child abuse involved 7.5 million children in the United States and that 3.2 million children received prevention and post-response services. The truth is that parents' and caretakers' role is to nurture, care, and provide safe and loving environments for the children. These children are under their care just for a little while, and children are not properties or assets; they don't belong to their parents or caretakers in any way. Their true parent is our Father-Mother-God where their souls originated from, and earthly parents' and caretakers' role is to guide, nurture, love, safeguard, and lead them during the first stage of their incarnations. The control, abuse, and poor treatment that some children face in this world must stop. This madness must stop. Let's all become aware of the truth that children are important and as valuable as any other adult, and let's honor, love, nurture, and care with tenderness for these amazing and wonderful beings that children are in truth.

There is so much to learn about children. Indeed, children came into the lives of their parents and caregivers to teach them something, and there is a lot to learn by spending time with children. Children are our leaders and our healers because they show us how to live life, how to live in the present moment. Indeed, contrary to adults, children do not live in their mind, in the egoic mind, but only live in the present moment; they live only in the now. Therefore, they do not hold grudges or carry resentments and all these toxic feelings that adults pack themselves with. Children are free, happy, and easy going; they forget easily and let go of offenses readily. "The Kingdom of Heaven belongs to those who are like these children" (Matthew 19:14). The kingdom of heaven is, of course, a metaphor that represents the peace, joy, love, and freedom that we all are seeking on Earth. To experience this heavenly state of peace, love, joy, and freedom, the ego must be put away, and we must become like children by rediscovering how to live in the present moment by becoming easy going. We must let go of hurt, old unforgiveness, anger, past regrets, stories about the past, and worries about the future and live only in the present moment, in joy, and be playful and happy as children do. We must become like children. This is why children are our teachers and our healers; they show us how to live. This is what it means when it is written is some religious scriptures: "The Kingdom of Heaven belongs to those who are like these children."

Indigo Children, Crystal Children

Is There Something Else That You Would Like to Tell Us about Children?

Yes, currently, the earth is changing and shifting into the fifth dimension, and a new generation of children is coming to populate the earth. Many of these new-generation children that are coming to the earth are the indigo children and the crystal children. These children are gifted, more evolved souls who are coming on Earth with specific and much-needed life purposes. However, unfortunately, these new generations of children who are populating the earth currently are not recognized for whom they truly are, and they are labeled as 'troublesome, weird' or misdiagnosed by healthcare professionals, parents, and their teachers as having 'behavioral issues.' For instance, indigo children are a breed of souls who are incarnating on Earth with the mission to bring about change in the educational, political, nutritional, and organizational aspects of the society, and their purpose is to create change by shaking off the old dust of dishonesty, ineffective educational curriculums, and the nonsense rules of our society. They were prepared to resist the old ways of doing and came, having decided not to fit in with these old rules. These children, therefore, won't comply and do not want to be forced to fit in or comply with society's rules since their mission is to challenge these rules and transform them. Unfortunately, these children get diagnosed as having ADHD (Attention Deficit Hyperactivity Disorder) or ADD (Attention Deficit Disorder) and are forced into submission by being medicated with drugs. The truth is that these children have no disorder, and they have no deficit of any kind; they are simply 'indigo children' who are merely the new-generation children that came prepared to resist the pre-established norms of our society. Therefore, they are naturally noncompliant with rules, whether they are at home or at school. Yes, they may not care that much about following directives at school and may have a lot of energy, but this is just because they are warrior-spirit tempered and are leaders by nature, and in truth, they have no disorder, and there is no need to drug these children and put them under powerful medications that will alter their brains and silence their gifts and natural talents. Many of these diagnosed ADHD and ADD children are, in fact, indigo children who are highly intuitive, telepathic; equipped with leadership abilities, with warrior spirit temper, and with life purposes to bring their gifts and talents to help the world to evolve. Indeed, these children are little lightworkers who came to assist and bless the earth with their unique gifts and talents. Unfortunately, they are drugged, forced to fit in and to submit to the outdated and limited societal norms and rules because they are misunderstood and because people do not know the truth of life and the truth about them.

The second group of this new generation of children who are being born currently on Earth are the 'crystal children.' Crystal children, in contrast to the indigo children, are easier going, even-tempered, highly intuitive with psychic ability, forgiving, highly empathetic, extremely creative with artistic abilities, gifted in many ways, sweet, and loving. Unfortunately, many of these crystal children are not understood by society either, and they are being labeled as 'weirdos,' and many of them are being diagnosed with autism or Autism Spectrum Disorder (ASD).

This is not to blame the healthcare practitioners, parents, teachers, or anyone but to simply shine the light of truth on what is going on and where we currently are in our evolution in this world. The healthcare professionals simply don't know because they have not been trained to understand beyond the physical bodies. They are merely following what they have been taught in their medical school curriculums. Many, including teachers and parents, simply falsely believe that there is nothing else beyond the physical body and that all problems must be dealt with using medications and drugs. The healthcare practitioners are merely following guidelines created by pharmaceutical industries and committees who are driven by money, profits, and greed. It is up to every single one of us to wake up, to learn the truth, take back our power, and to do what is best for our children and for ourselves. There is no need for blame, accusation, finger pointing, or feelings of victimization here. None of these have their place here, and none of these is the purpose of this book. The purpose of revealing the truth is simply to get more understanding so that we can start to live in a place of greater awareness and peace for all. Take your health and the health of your children into your own hands. Be proactive. Be curious. Do your own research. Seek the truth, and you will find it.

The Truth about Life
CHAPTER 13

THE TRUTH ABOUT MASS MEDIA, TELEVISION, AND CELEBRITY MAGAZINES

"If you want to have a healthy mind, you must feed your mind with truth. You must feed your mind not with junk or poison but with truth."

–Rick Warren

Everything in the world is vibration and has a unique signature frequency. For instance, if you are running outside or walking in nature or reading a good inspirational book, these are high-frequency activities, and you will start feeling good, energized, and happy while you are doing these activities and after the activities because these are the kinds of activities that will boost your vibration. This is the reason why many people feel good after going to the gym or walking outside in nature with their dogs. On the other hand, if you are sitting in your sofa, watching a violent movie or gossip shows regarding celebrity dramas, these are low-frequency programs that will pull you down and decrease your vibration and consequently will affect your mood and your health in the long run. Furthermore, by watching them, you are storing these low-frequency images and scenes in your subconscious mind, which will attract similar things into your life. These images, violent movies, will be running in the background of your subconscious mind, and this is one of the reasons why many people wake up in the middle of the night frightened with nightmares after watching violent or frightening movies.

What is entertaining about watching people kill each other? Would it be entertaining for you to watch someone kill your children or your loved ones? Why, then, in our society have we agreed to call movies of people killing others, gossiping magazines about tragedies, divorces, and sexual affairs of famous people "entertainment"? If your spouse cheated on you, would it be entertaining for you to watch everything displayed on "entertainment" shows on TV? Would you call this cheating "crispy news" or "the scoop of the day" that you would like to share with your colleagues, relatives, or the whole world for them to watch and give their unsolicited opinions and debate about them at lunchtime or in the office break rooms? Why, then, would you agree to "entertain" yourself with these kinds of low-vibration news, TV programs, shows, and movies? "Do to others as you would have them do to you" (Matthew 7:12).

Keep in mind that some of these celebrities are so affected by the information displayed about them on these shows and magazines that sometimes they live hidden away or even attempt to commit suicide because of the negative consequences these shows have on their lives—without even considering the detrimental impacts that those kinds of gossiping news and magazines have on their children, families, and loved ones. We have seen how far these types of negative news, shows, and magazines can go with the tragic death of the beloved Princess Diana of the United Kingdom. Why, then, do we keep participating in those low-frequency shows and magazines by supporting them with our money, by buying them, and with our energies by watching them and boosting their viewing statistics and audience numbers? There is no judgment here, and do not judge yourself or feel bad about yourself if you watch and enjoy these shows. The truth is that we have been conditioned in our society since childhood to watch these movies and shows. Instead, be glad that you now have the opportunity to discover the truth about the media system.

What about Informative News, Educational Programs, and Other Documentary Programs on Television Which Seem to Be Helpful?

Each of these subjects presented is only one of the various facets of the indoctrination system in place in our society. Keep in mind that whether you are watching a renovation show, a food show, the news, or even children's programs, there are frequent commercial breaks in most of these programs that you are absorbing in your subconscious mind. There are several subjects on television that may seem interesting, but overall, you may gain more by spending your time doing something more worthwhile such as working on a creative project, for example.

Moreover, by sitting and watching these programs, you may be filling your conscious mind with information that you don't necessarily need while at the same time promoting a sedentary lifestyle and obesity. For instance, as a pharmacist, I know the enormous impact of televisions shows and commercials on the population. How many times have we seen people rushing to pharmacies, requesting medications that they have seen on commercials or on health and medical shows? People watch the advertisements for a new drug on TV; it triggers them to ponder about their own health, and then they start to believe that the feelings and the discomforts that they are experiencing in their lives match the symptoms of the disease that the new drug promoted in the commercials and TV shows claims to alleviate. Consequently, they start to believe that they may have these diseases and later come to their local pharmacies, requesting these drugs promoted on television shows and commercials.

I will add here that the mass media system is so well tuned in and effective that there is a whole generation, even the global community, that now believe in the existence of 'flu season.' The truth is that the Universe does not create any season called 'the flu season' where many people will be likely to be sick and have the flu or certain diseases. The truth is that this is all a man-made and carefully planned idea. Sure enough, the population buys into this lie, and many people rush into pharmacies to get their flu vaccines done. The questions to ask are the following: Where does this 'flu season' suddenly pop up from? Why didn't it exist hundreds or thousands of years ago? Who are the beneficiaries of this 'flu season' invention? Were you dying from the flu several months of the year before some committees, industries, big pharmaceutical companies, or individuals came and claimed that some months of the year will be called the 'flu season'?

What Are Your Suggestions Regarding Television?

It is preferable to simply turn it off or considerably cut down the time spent watching it. What most people believe and the arguments that they give when it comes down to turning off the television is that if they turn their televisions off, they won't know what is happening in the world anymore and that they will be cut off from the world. However, this is not so; this is just the fear that the ego mind will come up with, convincing you that you will not know what is happening in the world once you turn your TV off. In fact, all the information that you really need to know in life will come to you somehow. For instance, you will go to work, and your coworkers will inform you of the big events that have happened in your country or in the world, and whatever you need to know will come to you one way or another either from your friends, relatives, coworkers, or even strangers. Therefore, you do not need to sit on your couch for hours, ingurgitating the negative news of television. The Universe is very intelligent; if there is something or some news that you need to know for your growth or welfare, the Universe will bring this information to you. The false belief that you will be cut off from the world by turning off your television is nothing else but false thinking, the fear of being odd, and an empty argument that keeps many people watching the news.

In fact, watching television can become a very addictive habit; therefore, it is normal that there will be some fears at the beginning when you decide to turn off the television. I have gone through this process myself, and I know all the fears and arguments that came to my mind once I made that decision to turn the television off. I was so afraid that I was going to be bored or that I would miss essential information if I stopped watching the news or that I would be odd and cut off from the world. However, my experience has proved to me that I lost nothing and instead gained considerably by making that decision, and I never regretted it. In fact, when you turn your television off, you will have more time in your schedule for activities such as meditation, creativity, and so on. What will happen is that once you turn your television off, you will also have more time to take care of yourself and to explore creative pursuits, and even the goals and dreams that you have for your life will start to become clearer in your mind.

Reducing the time spent watching TV will make space in your body-mind vessel for new good things that are of high frequency to come in your life. For instance, let's suppose that you are a vessel, a vessel like a vase, container, or a jar. If your vessel is filled with sludge, will your vessel have room for clear water to pour in? Of course not, because there will be no room in your vessel for this clear water to enter. Where will this clear water go if your vessel is filled with sludge? However, once you have cleared your vessel and removed the sludge, you will make room for the clear, pure water to flow in. This clear, pure water is just an analogy for things such as inspired ideas, creativity, abundance, high-vibration activities such as gardening, reading high-frequency spiritual books, going to watch artistic shows such as live ballet in your community, walking in nature, pursuing your true goals and interests in life, and so on.

I would like to mention here that many successful people in the world do not spend their time watching TV shows and programs, and some such as Oprah Winfrey have admitted that they barely watch TV because they do not want these negativities to fill their minds. In fact, many truly successful people do not squander their time watching television programs. I am not talking about success just in terms of money but in terms of happiness, peace, love, and true fulfillment. Many of these successful people cut down their television time or simply turn their TVs off and therefore do not carry these negativities in their minds, and that is why they are able to receive creative ideas, business ideas, and inspirations; they have made room in their own minds for creative ideas. In fact, they spend a lot of time dreaming about their goals, pondering on how they can improve their businesses or where they will find the next best business ideas and how to serve the world. On the other hand, the great majority of the population are feeding their minds with low-frequency TV programs or thinking about and discussing the last natural disasters or the last shootings that they have watched in the news. In truth, these truly successful people are well aware of the law of manifestation, the power of thoughts, and know that they cannot afford to pull down their energies with dramas and the low-vibration information of the television news and programs. These abundantly happy people, either consciously or unconsciously, are applying these concepts of the law of vibration in their lives by keeping their minds and thoughts focused on positivity, love, happiness, and high-frequency activities.

What Can You Tell Us about Celebrity Magazines?

It is evident that the information provided in these celebrity magazines is not of high vibration. Do you really need to know that a movie star is going to rehab or that a famous actress wore the prettiest or the ugliest dress for the Oscar ceremony? What good can such information bring into your life and what is it useful for? Of course, these celebrity magazines are low-frequency broadcasting systems that feed the population with dramas and make gossiping an acceptable practice in our society. It is only the ego who loves stories about people, dramas, scoops, and sensational news, and who finds drama entertaining. Why not simply mind our own business instead of trying to know what is happening in celebrities' lives or others' lives? These magazines make it seem like sneaking the nose into others' business and private life events is acceptable and worthy of praise so much that they call these practices 'news' or 'magazine articles' and publish them for sale.

Some people argue that reading this news about celebrities helps them not to feel so bad about themselves and about their lives to know that celebrities are going through the same life challenges as themselves. I would say that this type of thinking comes from the ego and that these justifying statements are similar to saying, for instance, that seeing others die from terminal diseases makes you feel better about your hypertension or diabetes. How would you feel if your neighbors told you that they were relieved to know that one of your children was using drugs too because it made them feel better to know that their child was not the only one addicted to street drugs? The truth is that there is nothing entertaining in reading about others' divorce, marriages, scandals, sexual affairs, etc., whether they are celebrities or not.

Whether we call a person a celebrity, famous, star, idol or not, keep in mind that he or she is someone's child, someone's sister, someone's father, someone's spouse. These drama-filled informational magazines are not the type of entertainment that your soul is longing for, and their underlying purpose is not based on love, expansion, growth, or evolution, which are the virtues that your soul is seeking, and this drama-based information will only pull down your vibration, whether you are aware of it or not.

What many do not understand is the energetic aspect attached to these types of magazines and media systems. For instance, every time journalists steal personal information from celebrities against their will and sell it, energy-wise, it is as though they are sucking energies out of these celebrities' lives to sell and make money out of it. For instance, if someone wants something from you, and you are very sensitive to energy, and your vibration is high enough, you can feel your energy being depleted in the presence of that person. Therefore, what these trashy magazines are doing is not just a harmless matter on a spiritual and energetic level. Buying these magazines and reading them means that you are indirectly participating in these types of unloving activities and the negative effects that they have on these famous people while dragging down your own vibration at the same time by feeding your mind with toxicities and unloving information. Therefore, it is

THE TRUTH ABOUT LIFE

preferable not to buy these magazines. I will even add not even to grab them and read them when you are at the lines, waiting to check out your shopping items in grocery stores. These magazines and TV programs continue to exist in our society because people who produce them argue that this is the type of information that the population wants to hear, see, and read. It is up to us, then, to decide that we do not want the trashy, the drama, the scandal, the unloving, or any low-vibration information. We must decide that it does not entertain us to hear or read about other tragedies, disasters, killing, and that we want to hear only loving things, that we want to wake up toward our divine nature, toward love, expansion, creativity, and joy by turning away from anything that is not of love.

The Truth about Life
CHAPTER 14

TRAGEDY, DISEASE, NATURAL DISASTERS, PEOPLE WITH DISABILITIES

"The man who fears no truth has nothing to fear from lies."

–Thomas Jefferson

Why Do People Die and Why Do They Die Sometimes in Tragic Ways?

Souls often plan the times when they would like to return to the nonphysical plane before their incarnations, and they often plan the lessons that they would like to work on Earth. When they feel like they have learned all the lessons and done all that they planned for their incarnations, they decide to go back home to heaven. Some plan to leave early as children, or as young adults, while others plan to live until their old age.

For instance, a black man may plan before his birth to return home to heaven in a tragic way such as in a gun shooting by a white policeman to shine the light on the phenomenon of discrimination so that these types of discrimination may be publicly shown to the world in order to be addressed and corrected. This soul (the young black man), in a way, has lovingly chosen to use his incarnation and his life as a service to help humanity to address the false beliefs and issues of superiority and inferiority that create so much suffering on Earth, while in truth we are all light beings with no race, color, or ethnicities. This may even be an agreement with the soul of the white policeman who has killed him and all the other souls involved who may have chosen to play these roles to help their nation and the world. I want to reiterate here that I am not condoning or justifying anything but simply shining the light on some possible ways that people may choose to die (return to the nonphysical plane). Keep in mind that no one really dies because we are immortal light beings. Also, we are all One and brothers and sisters in spiritual truth; therefore, sometimes people may choose to die in tragic ways such in a gun shooting by a policeman to bring awareness of some discriminative practices that used to be hidden before in order to help shake off the old destructive behaviors and help humanity progress toward love. My intention of mentioning this is to bring clarity and understanding to some tragic ways that people sometimes die (transition into the nonphysical world) that we do not understand. Thus, I call upon compassion, love, unity, and peace for us all.

Why Do Some People Die Despite All the Prayers of Their Loved Ones and Friends?

There are times when people may have prayed for a loved one who was sick or injured so that this loved one may live or be healed, but this loved one finally died, and they became really disappointed and believed that their prayers did not work. When that happens, they believe that their prayers are ignored or unanswered, and sometimes they become angry at God, stop praying, or even become atheist. The truth is that everyone has a free will choice, and when someone is sick or injured, God and their guardian angels show them on the spiritual plane different scenarios for how their lives can turn out if they stay alive and if they leave the physical world (die), and they are given the freedom to decide and choose from the different scenarios presented to them. They are shown what is going to happen if they stay alive and what the outcomes may be if they go back to heaven, and sometimes they decide to go back to heaven out of love because they don't want to become a burden on their families or for various other reasons. If they go back to heaven (die), their loved ones and families perceive that as a tragedy; they mourn, and they often get angry at God for not answering their prayers. However, what seems like a tragedy for them is simply a decision of the souls of our loved ones and the best for their spiritual journeys. This does not mean that their prayers were not answered but simply that the souls involved have decided that it would be best for them to return home to heaven.

Why Do We Have Earthquakes and Other Natural Disasters?

This is an interesting question. Many people wonder: *If God is good, or if there is even God, why does He allow disasters such as earthquakes to happen?* The truth is that we create with our thoughts, and we create from inside out. The exterior events and the natural disasters are simply manifestations of what is happening inside of the collective mind of humans. We are indeed powerful creators, and when a great number of people are holding unloving thoughts and have war-like thoughts in their minds toward their neighbors, coworkers, relatives, or all these angry thoughts about other nations or the government and so on, the result can be disastrous and turn into natural disasters that we are seeing in the world. This is because thoughts are not just thoughts, and angry, hateful, judgmental thoughts will lead to these external manifestations and attract events such as natural disasters. The opposite is true as well; holding loving thoughts and thoughts of peace and joy will lead to more global peace and world peace. Energy is like sonar; whatever energy you put forth will come back to you. Also, there is a meaning and a correlation to some of these events and natural disasters as well. For instance, places or locations where there is great air pollution or some sort of abuse of the air will more likely experience disasters related to the air such as tornados and so on. Everything is connected, and everyone is a contributor either to world peace or to the calamities via our thoughts and the energies that we are holding in our minds and feelings. The best thing that we all can do is to work on ourselves to hold thoughts of peace, love, and acceptance toward ourselves and everyone in our lives and at our workplaces.

As to the question "Why does God allow this?" the answer is simple: God does not allow anything, and God does not do anything to us or to anyone. We are the creators of our world, and every thought, every word, and every action contributes to the collective environment. God has given us free will and the power to create with our thoughts, and God will not interfere and take back the gift of free will that we have been given. As we can see, we can use the gift of free will to create beauty or to destroy our world. The choice is ours at any given moment. We all as a collective of humanity need to get out of the victim consciousness that we are in with the false beliefs that we hold that God is doing something to us or that it is God who is bringing natural disasters to us, or why God is allowing that, or God is punishing us, etc., and take back our power and take responsibility for our thoughts, feelings, words, and actions and create the peaceful world that we want to live in. What we all can do here is to be loving and to embrace the revolution of love that is currently happening on our planet. Love for ourselves, love of others, respect for our environment, peace of mind, and praying for world peace is the answer for our world.

Why Are Some People Born with Disabilities Such as Down Syndrome and Why Are Some Children Born and Only Experience Pain, Suffering, and Disease: Whose Fault Is That?

Health issues affecting children and the suffering of children is one of the most challenging things, and it can be very disturbing. Yet, it is no one's fault, and no one has sinned or done anything. It can be due to various things. Sometimes, there are experiences or situations where the children involved may have elected themselves before their births. It also can be an accord that these children have made with their parents for learning experiences on Earth. It also can be that the child or their parents have elected to work on some karma from past lifetimes. In truth, it can be due to many things. The most important thing to keep in mind is that no experience or birth condition can come to anyone without them choosing it somehow either before birth or consciously or unconsciously. We are here for learning experiences, and some experiences are learned through pain. Also, some elect to learn lessons about fear and suffering, and this third-dimensional plane provides the opportunity for such experiences. The good news is that we are eternal, and no one ever dies, and no matter what the earthly experience is, it is just a temporary experience.

In spiritual truth, no one is disabled. The soul is healthy. People simply may have elected to experience a particular life circumstance or physical condition. Sometimes, it is during these lifetimes that we learn a lot about compassion and what it feels like to be perceived differently by others or to be different from others. Keep in mind that an incarnation is mostly like roleplaying for learning opportunities. The souls who elected to incarnate in a position of disability sometimes are even more in touch with their divine nature and learn a great deal about compassion during these specific incarnations. The best thing to do is to have compassion for everyone, no matter the path that they have chosen for their life, and to remember that we are all souls equally important, no matter what we have elected during a specific incarnation. We are all teachers of one another and brother's keeper of one another.

The Truth about Life
CHAPTER 15

MISCARRIAGES, ABORTION, SUICIDE, DRUG ADDICTION

"Facts are many, but the truth is one."

–Rabindranath Tagore

What Can You Tell Us about Miscarriage and Abortion?

The truth about miscarriage and abortion is that souls choose their parents and the time that they desire to incarnate. It is agreements between all parties involved (parents and the child), and every one of them agrees with the terms of their mutual contract. Sometimes the souls of the child may choose to return to the spiritual realm, which means that they have changed their minds in a sort either because the birth condition is not suitable for them or because they feel like they are not ready anymore for the incarnation or for several other reasons. In these cases, the soul of the unborn child may decide to leave and return home to heaven via miscarriage or even via abortion. It is all done in accord with the child and the parents involved on the spiritual plane.

Miscarriage seems tragic in our human perception, and yet on a soul level, it is all done in accord with all souls involved. Keep in mind that we have free will choice, and every soul chooses to do whatever they deem best for their spiritual growth and evolution. Sometimes, it may just be the agreement that the souls had made or elected before birth and the experiences that they choose to work on. Even for abortion, it is a common agreement for all souls involved: the mother, the father, and the unborn fetus. Sometimes, the soul of the child may choose to come back later to the same parents, and other times, the soul may choose to incarnate with different parents after the miscarriage or the abortion. When a soul has decided to incarnate, it will find birth parents that will give the experience and the lessons that it wants to work on. There is truly an infinite number of possibilities, and there are billions of people on this planet to choose from as parents. Understanding this helps to place things into perspective and helps us not to judge ourselves or others. Raising a child is not an easy task, and sometimes either the mother or even the fetus decides to back up from that responsibility and change their minds. No one should then judge anyone for abortion when it is done with love because all is well at the end. Either the soul of the baby will come back when the time is more appropriate, or it will simply choose other parents. Parents who have experienced miscarriage(s) should know that the baby is not dead in any way and has simply chosen not to stay on Earth of its own accord.

What Can You Tell Us about Suicide?

Sometimes people decide to incarnate to have a particular experience, but the experience can become too much, too painful, or too dramatic for them once on Earth; therefore, sometimes they decide to leave and go back to heaven via the means of suicide. It is their souls' decision, and God respects that. It is more than we can comprehend or understand with our human knowledge, and it is tragic for us humans, but all we need to know is that all is well at the end and that no one truly dies. Death is simply a doorway to the nonphysical world of heaven just like birth is the doorway to the earthly life. All is well.

There has been misinformation, condemnation, and so many fears on this planet stating that people who commit suicide go directly to hell, but this not true. They do not. After their suicide, they are simply welcomed by God and their guardian angels to the Light like everyone else. Furthermore, hell does not exist, and everyone simply goes back to the heavenly afterlife plane where they review their lives and continue their existence in the nonphysical plane.

Also, there has been condemnation of people who commit suicide. Please, do not condemn them because you do not know what was going on in their lives and what pain and suffering that they had been going through to get to this place of suffering, total loss of hope, and despair. Have compassion on them and on their families. These families are already going through tremendous pain enough, and there is no need to add more pain to their suffering with statements and beliefs that their loved ones who committed suicide are going to hell or anything of that nature. Have compassion for everyone.

Why Are Some People Addicted or Suffer from Drug Addiction?

People elect the lessons that they want to learn. Sometimes, when a lesson is not assimilated, they may repeat these lessons or the experiences in future incarnations until they learn what they are supposed to learn. Sometimes people may suffer from addiction during lifetimes but are unable to overcome their addictions during their lifetimes. Therefore, they may repeat these lessons and come back and try again in future lifetimes until they overcome their addictions. These souls may elect to choose parents who are going to enable them to have the experiences that they desire. They may choose abusive parents or addicted parents and may become addicted later if their life lesson is to learn to overcome addictions later on in their life. All are learning experiences, and sometimes lessons are repeated until they are learned.

I want to mention here that people who are dealing with addiction are already in great suffering and simply don't know the best way to deal with their suffering; the addiction is like a way out for them, but of course, this is not the best way. An analogy to help to understand the situation of those dealing with addiction is the following: Let's suppose that a lady was drowning, and a rescuer went to rescue her; however, the rescuer realized that both of them were going to be drowned, as the lady was pulling him down. Not knowing what to do and how to save himself, the rescuer then decided to knock out the lady to save his own life out of survival. Of course, this is not necessarily the best way to deal with his situation, but he just wanted to save his own life and wanted a respite at all cost. Now, he has created another problem with his choice by knocking out the person who was drowning, and he will have to find a way to get the lady out of the water. However, temporarily, the rescuer opted for that solution because he simply wanted relief from his challenge and wanted to save his own life, and he did not want to be drowned.

This is the same situation that people dealing with addiction find themselves in. They become overwhelmed and do not know how to deal with whatever they are going through and use drugs and chemicals as coping mechanisms. Of course, this is not the best approach, as it only creates other issues for them, and they will have to eventually deal with the core issues. Also, if they fail to overcome their addictions, they may have to come back in another lifetime and try again and again until they succeed. I will call out here to compassion and the practice of the art of nonjudgmental behavior toward all people, including those who are dealing with addictions, because they are not in an easy place. They are already going through a great deal of pain and suffering, and they don't need more judgments, blame, criticisms, or rejections from their parents, society, or from anyone. What is needed here on Earth regarding these issues is compassion and unconditional love.

The Truth about Life

CHAPTER 16

FUTURE AND FATE OF THE EARTH, DIMENSIONAL SHIFT, AND GRADUATION

"Teach this triple truth to all: A generous heart, kind speech, and a life of service and compassion are the things which renew humanity."

–Buddha

Is the End of the World coming, and Are We in the Last Days?

No. The world is not going to end. The earth is here and will remain here as far as I know. However, the earth is evolving, and it is moving into higher dimensions: a fourth and fifth dimension. This means that the vibration of the earth is increasing. The whole process of evolution of the earth to a fourth/fifth dimension as well as the graduation is sometimes referred to as dimensional shift. It is an increase of vibration of the planet to a higher, more loved-filled and lighter energy. A fourth dimension is a dimension where there will be more love, peace, unity, and cooperation. The fourth dimension is less dense, more fluid, easier, and more loving. The evolution of the earth to a fourth dimension is what was misinterpreted in some spiritual scriptures as the end of the world. However, it is simply an evolution or progression. You can see this as graduation. However, it is a graduation where there is no passing or failing. Those who will not graduate are not failing anything. They will simply continue their evolution progress in another third-dimensional plane very similar to the earth. That is all. Because we are in a graduation period, there will be a separation of souls at the end of this incarnation. "Two men will be working together in the field; one will be taken, the other left" (Matthew 24:40). This graduation season has already started and will last for approximately a few hundred years. This is not to put pressure on anyone because all is well, and everybody will graduate one day either in this lifetime on Earth or somewhere else in another third-dimensional planet.

Can You Expand More on the Graduation or the Ascension Process?

Graduation is a passage from the third-dimension plane to a fourth/fifth-dimensional plane. Some refer to this as ascension. It is when you have learned all your lessons needed and have awakened from the illusion or dream of fear and separation and do not need to come back on Earth or go to another third-dimensional plane to continue your spiritual evolution. It is something that existed throughout time, and many people have already ascended such as Jesus, Enoch, Yogananda Paramahansa, the Buddha, and many others. We are in a special period of the earth evolution where more people are awakening up and are graduating. Some may ascend and leave their bodies, while the majority of the 'graduated ones' will die normally but nevertheless will ascend to a fourth or fifth dimension. The ascension process is already ongoing on Earth right now. This means that some of the new babies being born currently are of the fourth or fifth dimension, while others are of the third dimension. There is a mixture right now, but progressively, all who will come will be fourth and fifth-dimensional beings at the end of the process, as there is a process of sorting out going on, even though the expression 'sorting out' is not truly the most appropriate term for this process.

This is all good at the end, and everything is going to be well as we are progressing in this change, and the future of the earth is glorious.

What Should We Do in Order to Graduate?

There is no need to strain or worry about this because everyone will graduate one day either in this lifetime or another lifetime. All I can say is that in order to graduate or ascend, you will have to be loving and emanate love in your thoughts, words, and actions at least 51% of the time for all combined (thoughts, words, actions) and will have to live from this space of unconditional love toward yourself and everyone. I will say that this is a complex matter, as it takes into account several factors such as things done throughout past lives and one's evolution as a whole. The only thing that you can do is to do your best and be the best that you can be. It is a manner of vibration, so increasing your vibration and doing everything that will increase your vibration will help you in this. The more important aspect of this is to love: loving everyone, no matter who they are or what they do. Forgiveness is also an important part of this process as well, as it will lead you to love and accept others unconditionally. Letting go of judgments (judging and criticizing others) will also align you with love. Basically, move toward anything that can bring you closer to love: kindness, compassion, serving others, meditation, living in joy, doing what makes you smile or what brings you joy, etc., being careful and watchful of thoughts, letting go of all negative feelings such as anger, resentment, jealousy, bitterness, gossip, etc., and truly being aligned with love.

The best way of doing this is not to focus on the graduation process that is currently happening but simply do your best and enjoy your life. Otherwise, this will put unnecessary stress and fear in you. From my personal experience, I remember the first time I learned about the graduation; I freaked out, and I started thinking "*What percentage of my thoughts are loving and what percentage are negative?*" and I started to worry about whether or not I will graduate. Right afterward, when I saw people, I sometimes said in my thoughts, "*Do you know that the graduation process is going on?*" until I finally gave up and decided to live as best as I could. The most important thing to remember is that we are immortal souls and that everyone will graduate one day from the third dimension either in this lifetime or another. This is meant to be a joyful adventure and exploration for us.

Why Are You Saying Things Look Great, but When Listening to the News, It Seems Like Things Are Not Getting Better, and There Are Disasters Everywhere? It Looks Much More Like Things Are Getting Worse on Earth. Can You Explain This?

Of course, there are tragedies happening on Earth, but the news that we get from the media and on television is not an accurate reflection of reality. This is not a political debate, but as the question is asked, I will address it. The truth is that things are getting better, and there are many good things happening on Earth right now, but this is not information that you are going to hear in the news and in the media. For instance, when you listen to your local news, you will feel like there are gun shootings everywhere every day. However, this only affects a tiny minority of the population of the US. More than 99% of the population of the US does not wake up fearing that someone is going to shoot them. In fact, more than 99% of the US population is concerned about how they are going to pay their bills or mortgages, about the education of their children and how they are going to care for their daily needs. Therefore, the news is not an accurate reflection of the truth and of what is happening right now on Earth. Currently, there are people on Earth who are devoted to helping others, and they are doing amazing things. Some are creating jobs in Africa, educating children, healing people, doing extraordinary things, and changing the lives of others. However, this is not the type of information that you will hear in the news. In truth, things are getting better, more and more people are accessing health care and education in poor countries, and more and more people are awakening to love on Earth. Things are getting better in the US as well. The reality is that there is more cooperation among people in the United States than ever before. Things are changing for the best all across the world.

In the case of our country, the United States, even though things may not really be as we wish, we have to admit that we have come far along compared to what it used to be hundreds of years ago. We all can honestly admit that it is much better today in the United States than what it used to be during the time of slavery or the time of segregation. Of course, it is far from perfection, and we have a long way to go, but it is true that things are getting better in most cases. The media portray stories and events that seem to show that things are getting worse, but on a spiritual level, what is happening is a sort of 'purification' or 'purging.' Because there is light and more awareness, and more people are living in the light, the darkness cannot hide anymore. Therefore, during this time of dimensional shift and graduation, all the darkness and things that used to be hidden are coming into the light so that they can be addressed and transmuted. The internet and the digital technologies available on Earth nowadays are enabling things to come to light and be shown. When these events are shown on TV, oftentimes we get confused and believe that things are getting worse when this is not the case in reality.

My personal suggestion is to turn down the TV and to surround yourself with positive and uplifting news; books; and gentle, inspirational movies. This will positively affect your vibration and help you in various ways as well. We are in a great and important time in our history on Earth, and light is shining through, and things will get better and better.

The Truth about Life
CHAPTER 17

CLARIFICATION OF SOME BIG MISCONCEPTIONS AND LIES ABOUT LIFE

"To reach the end goal you and I must be willing to be straightforward with one another. We must be willing and prepared to stare at ourselves in the mirror and speak the truth."

–Mark Devro

What Is the Truth about Wars and the Idea of 'Serving Our Country'?

Anything that goes against love, unity, and free will is false. Nothing can be further away from the truth than portraying wars as 'good' and believing that wars are the means of resolving Earth problems or conflicts. However, these lies are so compelling and masked with the statements that make them so convincing and so believable such as 'serving our country,' 'honorable,' 'sacrifice,' 'duty,' 'for justice,' 'for maintaining peace on Earth,' and so on. Many people believe in them because these lies come from governments, from man-made rules that we call statutes and regulations and because they were written and officially accepted as laws of our countries, etc. These lies become so embedded in our minds and even glorified as "honorable, acts of courage, sacrifice, service," etc.

The truth is that love, unity, supportiveness, end of separation is the answer to all Earth's issues. War cannot solve wars. Wars cannot and will not lead to peace. I believe that the aftermath of wars speaks for itself. In fact, no one truly benefits from wars; the country being attacked does not benefit from it, and the country that is attacking does not benefit either, as there are losses on both sides. How many destroyed people and families will we have in our society before we understand the falseness and the destroying consequences of wars? How many of our brothers, fathers, spouses, and loved ones will die and fall in the grip of war before we understand that we are auto destroying ourselves? How many people will come back suffering from PTSD (Post Traumatic Stress Disorder) and various kinds of emotional and mental issues before we understand that wars are not the solutions for Earth problems? How many traumatized children we will have before we understand that wars are lies that have been inculcated in our minds by society and the rulers of our countries?

I would like to state here that there is no need to condemn or point fingers at anyone because those who went to war and who are still going today have simply believed the lies and only did what they were asked to do. I would also like to state here that the governments and those who are making these laws and inciting others to go to war do not know better. It is up to us to know the truth and to decide for ourselves and go with love. Please, have compassion and mercy toward yourself and for everyone and don't condemn or blame anyone.

Light of Truth

Since I was young, every time I heard praise and eulogy about wars and justifications of them, I intuitively knew that these were lies and false principles. Several years later, during my last year of pharmacy school, I had the privilege to work in a couple of federal department pharmacies. One of my rotations was in Air Force pharmacy and clinic; there, I worked with young, valiant people filled with vitality who were engaged in the army. These were really amazing young people filled with hope and with good intent in their hearts of serving their country and helping. I believe that our governments recruit the healthiest, bravest, most dynamic, and smartest young people among the population as testified by the very strict physical and mental tests that they take before being recruited.

A few months later, after doing my rotation in the Air Force, I had also the honor to go to another federal hospital, a veterans' hospital this time, and I was devastated by what I saw and the pain and the suffering that many of our veterans were going through. I was deeply touched by the health, financial, psychological, and mental issues that they were dealing with. How had these young people filled with vitality, joy, health, and hope that I met in my previous rotation in the Air Force turned years later into these veterans suffering from PTSD and various kinds of health and emotional issues and struggling financially? This was the question that I asked myself, and I was saddened by what I saw.

I believe that we all, as a society and individually, should seriously reevaluate our principles and use our intuition and God-given power to know for ourselves the truth about wars and not buy into the lies that political leaders are telling us. How can a war be a good thing? How can killing others solve anything? How can transforming young, valiant people into veterans dealing with PTSD, depression, and other health issues be a good thing? Where do these false principles come from? How can we believe that engaging in a system where you freely give up your 'free will' and have to follow the commandments of others and even comply with orders to kill others can be a good thing? How can being constrained to execute orders that you may not agree with be a good thing? What kind of job entails that once you sign up for it, you cannot back down or change your mind—otherwise you will be considered a deserter, a traitor and will be considered to be doing something unlawful? I will say that anything that does not respect the principles of freedom and anything that will take away your free will is a lie. That is how you know what is the truth and what is a lie. Laws that take away your power; laws that are restrictive, dictatorial; laws that take away free will are lies and away from the truth. This applies not only to governmental laws but also to anything encountered in life. Let's be truth. Let's be love. Let's be peace.

What Is the Difference between Our Free Will Choices and God's Will (Thy Will)? Are Our Lives Predetermined by Destiny? How Do These Two (God's Will and Our Free Will) Interplay and Coexist?

God has given us free will, but at the same time, there is God's Will, which is shaping and directing our lives. Both exist at the same time. There are milestone events that come our ways in life that are designed by 'Thy Will' or elected before birth (by ourselves). However, between these milestone events, we have the free will choices on how to manage and react to these experiences.

To have a better understanding of this, you can see life like a river. 'Thy Will' is the flow of that river, and our free will is like paddles that God has given us to navigate in this river. With this analogy, we have freedom while we are in the river to navigate to the right, left, go forward in the river and yet at the same time, we are in this 'big river of life' that is taking its course and with a powerful current flow that is much bigger than our paddles. Trying to go against the flow of the river will be difficult, challenging, and painful. However, going with the flow and using the paddles that we have been given to navigate and make our ways through the river is much easier, smoother, and enjoyable. Yes, we have free will choices and, yes, at the same time, there is 'Thy Will,' which is shaping and directing our experiences.

It is better to see 'Thy Will' as elected experiences and events that you (your soul) have chosen before birth, before its incarnation, rather than the Will of a God who has chosen or predestined you to something. The truth is, all the milestone experiences and events that will happen in your life and that may seem like you do not have any control of have been chosen by you, by your soul. However, these plans or purposes happen to be God's Will for you as well. You see, your true free will choice (the true desire of your soul) is identical to God's Will as well. Let's take the example of Jesus, for instance. His soul elected before His birth to come to serve and teach, and then He would be sacrificed on the cross at the end. Let's say that in the midst of the difficulty and the pain ahead, He started to complain and said, "God, why have you done this to me? Why are they trying to kill me? Why should I be sacrificed on a cross? What have I done wrong?" Now, as shocking as it may seem to us, it is God's Will for Him to be sacrificed on the Cross because

God can see the future and knows that the outcome of His sacrifice will be good at the end and knows how many people will find hope and healing through His sacrifice. At the same time, it was Jesus' true Will as well since His soul chose to do that, since it was what His soul elected to do. Therefore, yes, 'Thy Will' exists, and at the same time, we have free will choice, and at the end, we can say that we have chosen all the experiences with our free will either before birth or after birth.

What Does Grace Truly Mean and What Is Grace?

When we think about Grace, most of us often think about unexpected favor of God or intervention of God to give us a breakthrough and jumpstart and something wonderful. These are all true, but grace is much deeper than that. Grace is the hands of God shaping humanity and the human experience. Grace is God's hands intervening to steer the creation and our human experiences in a particular direction.

In this way, grace can also block or stop someone from going in a certain direction that is not aligned with God's Will (Thy Will), which in spiritual truth is the same as that person's true will (the person's Higher Self's free Will). For instance, let's take the case of an imaginary person that I will name John, and let's say that God's Will for John (which is the same as John's elected life purpose before birth) is to be a teacher, a coach, and a person who teaches, supports, and helps others. However, John's father wanted him to become a professional athlete because it was something that he wanted to be himself when he was young and hoped to see this dream fulfilled by his son John. At the same time, John also desires to become a professional athlete as well for the financial benefit that it provides, the social recognition status, and for several other reasons. However, even though John was a very good athlete in high school, was one of the best on his team, and had a great chance of being selected for a well-recognized university, an unexpected accident happened; John had an injury and could not perform at a professional level anymore. This was a tragedy, and it looked as though the world was against John because this injury stopped him in his ambition and aspirations. However, years later, John became a coach, and he taught and helped students, and he became one of the best coaches in his state and finally had a long, fulfilling and exciting career helping countless athletes to thrive and succeed. This accident or injury that John had may seem like a tragedy at the beginning, but on a spiritual level, it was Grace. It was the hands of God that stopped John in his ambition and his father's pressure and led him to his passionate work and true life purpose, which was to be a leader, a coach, and a teacher. It was all Grace. Yes, Grace can help propel you in the direction of God's Will or Grace can stop you in order to redirect you toward God's plans for your life. Either way, it is all Grace because the outcome at the end will be fulfilling for your soul.

What Can You Tell Us Regarding Creativity?

Another misconception that I used to believe myself was the idea that some people are creative and others are not. However, this is not so. When it is written that we are made in the image and likeness of God, this does not imply that God is a man or that we physically look like Him, even though it is not totally false since God can take any physical form that He or She wants and since we are all God as well. However, what this intrinsically means is that as offspring of God, we have inherited His ingenuity, essence, and creativity. This means that we have the same potential as God and that we are truly limitless in our ability to create. We have the potential and the God-given ability to create, to imagine, and to invent whatever we desire. Everyone, and I mean every single one of us, is creative and equipped with some creative gifts from God. When people think about creativity, they often think about writing, or painting, or visual or auditory arts such as singing and so on. However, creativity encompasses a variety of fields that go from cooking to designing automobiles, blogging, coming up with ideas for a project at work, and so on. I will say that there are an infinite number of possibilities as well, and everyone has their uniqueness and talents in a specific domain of creativity. The key is to discover which gifts and talents you have been given and what is naturally suited for you and your life purpose. Again, there are simple ways to discover this and following your passions or your natural interests will lead to that.

Furthermore, creativity is one of the assignments that are part of our plans when we incarnate. It is part of our life purposes to express our creativity and to create various things while we are in the physical body. What often happens is that many believe that they are not creative or that they don't have time for creativity, and when we come on Earth, we are caught up in worldly affairs with a stressful life, work, heavy schedules, and so on. Furthermore, modern society is organized in such a way that many simply come back from their workplaces and hang around on their couches to watch TV or fill the rest of their time with several other distractions. Cutting down these distraction and time spent watching TV will make time in your schedule for creative expression.

Also, many are discouraged when they are contemplating pursuing their creativity because they believe that they don't have enough time or think that it will take too long for them to reach their destination or that it is not within their reach. However, creativity is a one-step-at-a-time process, and the mountains appear to be long and arduous until you start to climb them one step at a time. For instance, for people who desire to write a book, if they write one page a day, in a year, they will have written a book of 365 pages! Yes, it can be that simple. The key is to take a little bit of time every day to work on your priorities even if it is just 15 minutes each day. It only takes a little bit of discipline and commitment. There is joy in starting a new creative project, and there is even greater joy in seeing your project successfully completed. Yes, we are all creative and all have seeds of creativity within us ready to be discovered and explored.

What Can You Tell Us about DNA and the Body?

The current understanding of science is that we have 2 strands of DNA with exons, which are the active coding parts of the DNA, and introns, which are like 'junk DNA.' However, this is not quite accurate. In truth, we have 12 strands of DNA, and even if only 2 strands are active, the other 10 strands exist in an inactive state and simply cannot be seen by our current scientific equipment.

Furthermore, the 'introns' that are thought by scientists to be just 'junk DNA' are not junk at all. Scientists have started to discover that truth, and many have started to be interested in these other parts of the DNA (introns) that were once thought to be junk or inactive.

The truth is that as we are moving toward the fourth dimension, more parts of the DNA will become active, and we will become what some refer to as 'purified humans' or 'upgraded human beings,' which are simply terms to describe this state of living at the highest level of vibration of joy and love.

In fact, even though the majority of people on the planet live only with 2 strands of their DNA active, there are few people on this planet who currently live with 6, 8, or even 12 of their strands of DNA active. This was also the case of beings who became enlightened such as Jesus, and that is why they were able to do all the things that they did: healing people, communicating with angels, performing miracles, working with energy, and so on.

Also, DNA and the genetic makeup of the body are malleable and changeable according to the thoughts that we are holding in our vibration. I believe that it is in constant change, and it is a dynamic process as with many other things in life. Scientists have started to discover this and call it 'epigenetic.' For instance, the neurons in our brains are in constant modulation, change, and evolution. Scientists are now exploring the concept of 'neuroplasticity' and have started to understand that the neurons of the brain can regenerate and form new neurocellular pathways. This is true not only for the neurons in the brain but for all the cells of the body as well. It is amazing what the body is capable of, and it is capable of self-healing and self-repair if proper actions and principles are applied.

In fact, it all has to do with vibration, and there is a way of increasing one's own vibration and activating all 12 strands of the DNA. Living in love, thinking loving thoughts, becoming vegan, staying away from electronics, fasting once a week for 24 hours to remove chemicals from your body, bathing in sunlight daily or as much as you can to fill yourself up with the rays of the sun, going outdoors or into nature daily, drinking a lot of water, purifying your water to remove the chemicals, eating only organic vegan foods, practicing daily meditation, clearing your chakra, praying, living in truth (telling the truth at all times and being true to yourself and your soul's honest voice), laughing, being in touch with your divine nature, and ultimately living in a space of love and joy are some of the tools that can help to activate the 12 strands of the DNA. Implementing these ideas will also lead to greater health as well as other benefits such as the faster manifestation of abundance and youthfulness.

THE TRUTH ABOUT LIFE

What Is the Truth about Aging?

The body is malleable according to the thoughts bestowed upon it. Aging is a concept, and it is affected by our belief systems. Most people believe that they will age by a certain age, they will become ill, and then they will die. Therefore, their beliefs become true for them, and that is what they experience at the end. However, the opposite is true as well. If you believe that you are young and that you will be forever young, consequently, you will remain young. Keep in mind that we are ageless as souls. In spiritual truth, the body is made to last forever. As an example, I personally started a series of affirmations that I listen to every day to retrain and rewire my belief system. Two of my affirmations are: "I am 23 years old" and "I am forever young." I was surprised to notice later that I look very young and never age. More surprisingly, people often ask me, "Are you 23 years old?" This is just to say that aging is truly a concept and a belief that can be changed and that the body is truly malleable according to the thoughts bestowed upon it.

Many people are realizing today that they can live up to 200 years, but the spiritual truth is that the body is made to last forever. However, the body is made to thrive in an environment of love and in a high, positive vibration. Therefore, when we are in a positive environment surrounded by love and when we are thinking loving thoughts, the body naturally radiates health and is more likely to stay young.

Everything is a matter of vibration, and such is the case with the body as well. The more you increase your vibration, the healthier you will be and the younger your body will become. Every food has a specific vibration, and every thought has a specific vibration as well. The spiritual truth is that as the vibration of the planet is increasing, there will be a time when our bodies will stay healthy and young forever.

However, we are not there yet, so, for now, everyone should take responsibility for their health and body by feeding their bodies with nourishing healthy foods and beverages and do their best to think loving and positive thoughts all day long.

The Truth about Life

CHAPTER 18

ANGELS, ARCHANGELS, ASCENDED MASTERS

"Truth will always be truth, regardless of lack of understanding, disbelief or ignorance."

–W. Clement Stone

Do Angels Exist?

Yes, they do. Everyone, no matter their religious background or belief, has at least two guardian angels assigned to them before their birth. These guardian angels stay with them during their entire lives. It does not matter if you are a Christian, a Buddhist, a Muslim, or a Hindu, if you believe in angels or not, or if you believe in God or don't believe in God, everyone has at least two guardian angels assigned to them. Our guardian angels are continually around us even if we cannot see or hear them. They help with various needs, but you must first ask for their help before they can intervene because they respect free will.

We also sometimes have deceased loved ones in heaven with us. For instance, someone may have two loved ones in heaven and two guardian angels around him at all times. Sometimes we have more than two angels around us, but it really does not matter how many angels you have because it is not a static thing, as the number can change upon your request or depending on your need. For instance, during challenging times, God may send more angels to you to help you, or you may ask for more angels to be by your side as well. We also have other beings such as Jesus, the Buddha, or other ascended masters who continually assist and help all who call on them. What is amazing about this is that we can also ask for more angels and more support at any time of our lives, and this request is always granted, but many are unaware of this gift and this great resource available to us. In fact, I recommend to everyone to ask for more help if you are going through tough times or struggling in life because these angels can truly make a difference in your life, and you will see a shift in your condition.

Because of the variety of beings that can be around us, many people prefer using the term 'guides' or 'spirit guides' to refer to the beings who are assigned to us. We are assigned these guides or guardian angels because, in fact, we need help and guidance while we are in physical forms. For instance, someone may have one deceased loved one and one guardian angel assigned to him. Let's suppose that this person is a Christian, for instance, and that at some point in his life, he is going through challenges and prays to Jesus and requests Jesus' help. Jesus will come temporarily to comfort him, heal him, and assist him with whatever he needs help with. Now, let's suppose that this person is not a Christian but believes in the existence of God and simply asks for God's help for healing while he is going through challenging times in life. If that person does not request specifically any religious figure and simply asks for God to help him, then the most appropriate angelic being will be sent to help him. Archangel Raphael, the healing angel, may come to help him heal, or his own guardian angels that are around him all the time may help him to heal regardless of whether or not this person is a Muslim, Buddhist, or whatever. Because of the simple fact that he requests assistance and asks for God's help, help is sent to him. Therefore, I will say to ask for help. Ask, and it shall be given.

How Many Angels Do We Have?

Everybody has at least two guardian angels assigned to them at birth. Sometimes we also have deceased loved ones around us, but you can ask for as many angels as you want to be around you.

What Are the Roles of Our Guardian Angels and What Do They Do Around Us All the Time?

Their roles are varied. They protect us, and they watch over us. They help us to be safe and peaceful while we are fulfilling our missions and learning our lessons. They also help to answer our prayers and perform various tasks such as helping us to sleep and with anything that can bring us peace. The only thing is that they respect free will and will only intervene if we ask for their help. The only time that they will intervene directly is when you are in a dangerous situation, and your life is threatened before the time that you have scheduled before your birth to go back to heaven. Even there, they will need your agreement and consensus on a spiritual level before they can intervene.

Are There Different Types of Angels?

Yes. There are various types of angels. There are guardian angels that are assigned to each and every one at birth who stay with us through our lives. There are specialized angels called archangels who perform specific specialized duties and who can intervene and temporarily help us when we call upon them. There are several other groups of angels. For instance, there is a group of angels called the 'band of mercy' angels who work closely with Archangel Michael for protection. There are also healing angels who work with Archangel Raphael as well as a group of angels called 'grace' angels who bring small miracles on Earth. I believe that there are countless numbers of angels and groups of angels; their number may be close to infinite, as the universe itself is infinite. The most important thing to know is that all the angels and the heavenly beings work as a team to help us and that they are readily available to help everyone.

Archangels

Archangels are specialized angels. The word 'arch' before their names means 'bridge,' as they serve as a bridge between us and our Source/God. The Archangels are non-denominational, unconditionally loving beings who help everyone regardless of their religious background or beliefs. They are unlimited and can be with many people, simultaneously having a unique and individualized experience with each one at the same time. Therefore, there is no need to worry that you are bothering the angels because they can be with an unlimited number of people at the same time. They are pure light and love, and their mission is to assist with God's Will of peace on Earth one person at a time. They have very specialized functions, but they often work as a team to assist us. The Archangels with whom I work are the 12 Archangels; their number is just symbolic. In truth, there are many Archangels created by God known or unknown to us.

Archangels are very powerful beings and can be seen as 'higher rank' angels who oversee the other angels; however, it is better not to see them as above the guardian angels because they all work together as a team to assist us. In fact, in the realm Divine, which is the only true realm that exists by the way since we live in an illusionary world here, there is only Oneness. For instance, a body needs all its organs, and the role of the heart is not superior or better than the function of the brain or of the lungs; similarly, a specific type of angels is not superior or inferior to the others, as they are all extensions of God to guide and assist us during our earthly journeys.

The Archangels are mentioned in many religious scriptures; notably, the name of Archangel Michael was mentioned in the Bible, and he is known for his protective roles. Also, Archangel Gabriel was mentioned in the Bible as the angel who announced the world-changing birth of Baby Jesus to Mother Mary. Furthermore, the name of Archangel Raphael, the healing angel, was mentioned in the book of Tobit, which is a scripture that is part of the Catholic-Orthodox biblical canon. The name of Archangel Uriel was also mentioned in other religious scriptures. However, many think that these archangels existed a long time ago and have intervened in the history of the earth and now live somewhere far away in heaven with God. The truth is that these angels still exist and are alive and are in our everyday life. They are not somewhere far away in heaven, but they are right here among us on Earth and are in our everyday life performing various functions and helping whoever calls on them regardless of their religious denominations. Below are a few of the archangels with a short description of what they do, their signature light colors, when to call on them, and a prayer to invoke their assistance.

DR. LILLY KOUTCHO

Archangel Michael

Archangel Michael is 'The angel of protection.' His name means 'He who is like God.' He carries a sword of light and a shield. He is the patron saint of police. His role is to ensure that everyone is safe. His colors are deep blue, sapphire blue, and royal purple. He clears energies and fears and cuts etheric cords of attachments. He also helps people to have confidence, strength, and courage. He also helps people to increase their faith when they need a boost in faith. Archangel Michael knows all the details of our lives' purposes, thus he also helps with clarity with life's purpose. Therefore, he is the angel to call upon when you want to have clarity about your life purpose.

Archangel Michael is well known for his role of protection as illustrated by the true story of a man that I will refer to here by the name of Eduardo to protect his anonymity. In fact, one day, Eduardo went on vacation with his daughter in Costa Rica, Mexico. Costa Rica beaches are among the best beaches of the Pacific Ocean and are known to be calm, warm, and very agreeable. It was not the first time that Eduardo went on vacation in Costa Rica, and he was really familiar with the sea in that area. While swimming in the sea, Eduardo ventured to go further away from the shore, and before he knew it, he was too far away and was battling to come back to the shore. He was trying to swim to come back and felt like his efforts were in vain and started to give up, as he was so tired and out of strength. He thought that day might be the last day of his life, as he felt like he was drowning and giving up. He decided out of despair to call for help; with the little strength that he had left, he started to say, "Help, help, help." He stated later that he did not know how, but an ocean wave came and propelled him further and carried him near the shore, followed by another wave, then another wave until he got to the shore. He did not understand how he ended up on the shore and how the situation turned around for him, but he was so glad to be alive.

Eduardo told the story of his miraculous survival and what happened to him on that day to one of his friends who happened to have some intuitive and psychic abilities. His friend tuned in and said, "Oh, I think I see what happened that day; it seems like when you were out of strength, you started to call for help, and an archangel was sent to rescue you and was pushing the waves of the sea forward to carry you to the shore." It was unbelievable, but yes, an archangel was sent to Eduardo to rescue him as he was praying for help. I believe it was Archangel Michael, the protector angel and the rescuer angel, who came to assist Eduardo that day and saved his life. What would have happened if he hadn't asked for help, I really don't know, but I do know that the archangels always respect free will choices and will intervene only when people call for their help.

There are thousands of other reported stories of people who have called on Archangel Michael's help and who reported being rescued and helped by him. For instance, a person in a frightening situation reported that after calling for Archangel's Michael help, surprisingly a police car was passing by and helped her to feel safe. Others talk about calling for Archangel Michael's protection when they lost control of the wheel of their cars on the road and miraculously escaping inevitable car accidents. There are countless other reported cases of people who have called for the help of Archangel Michael and who stated that strangers came out of nowhere to rescue them, and before they could say "Thank you," the strangers disappeared. In fact, Archangel Michael, as well as the other archangels, can sometimes take temporary human form to rescue people. Indeed, in extreme cases, when there is a need, and someone is in danger and asks for help, Archangel Michael can take temporary human form to rescue people and then disappear. The people rescued have often said that a stranger came or stopped by to help them, but the stranger left the scene quickly after helping.

Regarding Archangel Michael, Are There Many Archangel Michaels Helping Several People, or Is There Only One Archangel Michael Helping Everyone at the Same Time?

There is only one Archangel Michael helping several people at the same time. Archangel Michael, like God, is unlimited and omnipresent and can help an unlimited number of people at the same time and have a unique experience with each one of them. He is not constrained by time or space and can help all who call on him. Therefore, there is no need to worry that your concern or request is not important or too small or too big or that you are bothering him because he is unlimited, nonphysical, and omnipresent. This applies to all the other archangels and Ascended Masters as well. There is only one Archangel Gabriel and only one Archangel Raphael and so on. In the same way, there is only one Jesus who is working and helping countless people in the USA, Africa, Europe, India, etc.

<u>When to call upon Archangel Michael</u>

When you need protection

When you are in fear and need help to release your fears

When you need strength, courage, and confidence

When you need energy clearing

When your faith is low, and you need a boost in your faith

When you are afraid and need help or rescue; you may think of him as the 'Policeman Angel'

When you need clarity about your life purpose

If you are going through challenging times such as legal court issues and need justice to be served

Prayer to invoke Archangel Michael's help:

Dear Archangel Michael, thank you for surrounding me with your protective light and for keeping me safe.

Archangel Gabriel

Archangel Gabriel is mostly known as the angel who announced the world-changing birth of Baby Jesus to Mother Mary. Her name means 'The strength of God.' She is 'The Messenger Angel' and often portrayed carrying a trumpet. As the messenger angel, Archangel Gabriel delivers messages of God to people, and she also helps those who are messengers of any sort such as writers and speakers with clear and harmonious communications. However, her roles are varied, including helping children. She has a motherly and nurturing energy and is the angel who helps children and anything related to children. She often works in tandem with Mother Mary to assist in the welfare of children on our planet. She is also the angel who helps us to nurture our inner child.

Another duty of Archangel Gabriel is to help people with creative projects such as writing, painting, and anything related to creativity, and she can help everyone who calls upon her with their creative projects because she is unlimited. She also helps to encourage and motivate people in their creative works by channeling creative ideas to them. She often works in collaboration with Archangel Uriel to give creative insights and ideas to people who ask for their help.

Seeing white lilies or roses is often a sign of the presence of Archangel Gabriel with you.

<u>When to call upon Archangel Gabriel</u>

If you need help to raise your children or need help to conceive or to adopt a child

If you are going through a challenging time and need strength

THE TRUTH ABOUT LIFE

If you are going through a challenging time, and you need strength; her name means 'strength of God' for a good reason

When you need help with communication

If you are a messenger of any sort such as a writer or a speaker

When you need help with creative projects or when you need creative ideas

Prayer to invoke Archangel Gabriel's help:

Dear Archangel Gabriel, please come to me now and give me strength.

Archangel Raphael

Archangel Raphael is 'The healing Angel' who helps everyone in their healing journeys. His name means 'God Heals' because he brings the merciful healing love of God to all who call on him. His color is emerald green. This angel often appears as calm, gentle, and relaxed. Indeed, it is healthy to be relaxed, and he illustrates this well by his peacefulness. His healing light is an emerald green light, and it is with this emerald green light that He does healing work. This healing light is a gentle energy that can be felt as tingling or subtle waves. When you ask for his help, be assured that he will assist in all ways. He helps with anything regarding healing such as physical healing, emotional healing, healing of animals; or if you are in the healing career, he can also assist you to help heal your clients and patients; or if you want to start a healing career, he can also help with that.

He is also the Matchmaking angel who assists people to find their soulmates if they ask for his help. Furthermore, he is the patron angel of traveling and can help you to have a safe and enjoyable trip if you are traveling.

I have personally experienced the miraculous healing of Archangel Raphael in many situations and also with my healing work and know the benefit of calling upon him for assistance. For instance, I talk about the angels to one of my friends who I will call Carla. One day, she related the extraordinary experience that happened to her. In fact, Carla reported that one evening, her 12-year-old daughter suddenly came to her crying and stating that she had serious pain in her head and that her head was hurting. Her daughter was in so much pain that they needed to go to the hospital that night. Suddenly, Carla said that she remembered the angels and started to pray and asked her daughter to pray too as they were preparing and getting ready to go to the hospital. Carla said that a few minutes later, her daughter stopped crying, and she was surprised that the crying of her daughter had stopped, so she went to check out what was going on and saw her daughter seated on the floor of their kitchen, holding a pack of ice over her head. A few minutes later, her daughter was completely calm and went to sleep because she was completely healed, and there was no need to go to the hospital. Her daughter told her later that as she was praying and asking for the help of the angel, she received an idea to go the kitchen and to place a pack of ice over her head at the place where she was hurting. Carla and her daughter stated later that they knew that it was miraculous healing and intervention of the angel.

I have also heard the story of another person who had chronic kidney disease and who asked for the help of Archangel Raphael. Against all odds, this person, who was on a donor list, somehow experienced a turnover of her situation and miraculously found a donor match, and the transplant went well, and everything went well for her. Why did Archangel Raphael help her to find a donor, while, in some cases, he directly intervenes to heal? I don't have a clear answer for that. However, I do know that if you call for his help, he will intervene, and he will guide you to the best healing path and option that is best suited for your situation.

In fact, there are countless reported cases of people healed by Archangel Raphael in our world today. Actually, Jesus, during His incarnation, was in communication with the archangels, and Archangel Raphael, along with other healing angels and the heavenly team were the ones conducting the miraculous healing performed by Jesus. Jesus even clearly stated that He was not the one who was doing the healing and the miracles, but it was God. In fact, the miracle of the healing of the sick at the pool of the Bethesda written in John 5: 1–15 was attributed to Archangel Raphael. I will add that all the miracles conducted by Jesus were probably done by the heavenly healing team that was around Him.

<u>When to call upon Archangel Raphael</u>

When you need healing for yourself or for your loved one or your pet

When you need help to find the best healthcare provider for your situation or the best healing option

If you are in the healing careers such as physicians, nurses, pharmacists, physical therapists, counselors, etc., and need help with your work

If you are traveling and need a safe and enjoyable trip

If you are looking to find your soulmate or a romantic partner

<u>Prayer to invoke Archangel Raphael's help:</u>

Thank you, Archangel Raphael, for surrounding me with your emerald green light and for supporting me on my healing journey.

THE TRUTH ABOUT LIFE

Archangel Chamuel

Archangel Chamuel is 'the finding angel.' His name means 'the eyes of God or the vision of God.' He is the angel who helps people to find what they are looking for, whether it is a missing object such as a key, a lost pet, or anything that you may be looking for and searching in life. His color is ruby red, tending into deep pink. Whether you are looking for a missing object, the right place to live, a soulmate, the right career, or anything that you feel is missing in your life, and you would like to find it, Archangel Chamuel is the perfect angel to call on for assistance. Nothing is ever too small or too big for Archangel Chamuel to find. No matter what you are looking for, know that by simply asking for his assistance, he will help you to find it. His answer may come in various ways, but it will always come in the perfect scenario and timing for you. Archangel Chamuel is proven to bring in the midst of all who ask for his help the things, the objects, the people, or the situations that they are seeking. Trust and relax knowing that nothing in this world can be hidden from this angel's extraordinarily powerful and loving vision. With his powerful vision, he can see where everyone and everything is in this world.

I have been amazed by the efficiency of Archangel Chamuel in finding objects and many other things, and with my personal experience, I came to completely trust his ability to find things and I do not hesitate to ask for his help. I remembered one day when the pair of flat shoes that I usually wear at home, which were made in Africa, was broken, and I was looking for another one. I went to an African store in the small town of Midwest USA where I was living at the time and could not find it. I searched on the internet and Amazon, with the hope that

I may find something similar, but my search was in vain, and I couldn't find anything close to the type of shoes that I was looking for. I was very disappointed and was considering ordering them from Africa, but it would probably take months before I would get them. Then I remembered that I could ask Archangel Chamuel to help me to find these shoes, so I did. I continued to search for them on the internet, and then I gave up and even ended up forgetting about them. A few weeks later, one day, out of nowhere, my sister who lived in California and who knew that I was looking for these shoes just sent me a text message on my phone, telling me that she was doing shopping somewhere in California and found the shoes and bought two pairs of them for me and that she would ship them to me via the mail soon. I was astonished and couldn't even believe it, and then I remembered that I had asked for Archangel Chamuel's assistance. I was very grateful.

I remember another time when I was at work, and one of the pharmacists with whom I working was looking for a drug to make an intravenous infusion for baby TPN (Total Parenteral Nutrition) and could not find it. She was looking everywhere and asked for help, and there were three people searching for this drug everywhere in the pharmacy and could not find it. Knowing that Archangel Chamuel had helped me several times in the past to locate missing objects, I mentally said a prayer to ask for his help, and just after asking for his help, an insight came to me to text another pharmacist who was off that day and who was normally in charge of making the baby TPNs. I quickly followed through with that guidance, knowing that it was probably coming from Archangel Chamuel. A few minutes later, the other pharmacist that I texted replied back to my text message and told us that there was a mistake in the physician's order for the TPN and that our pharmacy does not stock the drug we were looking for and that drug has to be substituted (replaced) by another drug. I was so amazed and happy that we had finally solved this issue and found a solution regarding the appropriate drug to use that day for the baby TPN and thanked the angel for his assistance.

I couldn't even tell how many things this angel has helped me to find such as missing objects, people, and he even helped me to find the perfect business partners.

Another time, during a radio interview, I was asked to talk about the archangels, and I briefly mentioned the name of Archangel Chamuel and his specialty, which is to help people to find what they are looking for. I had the testimony later of the story of a lady who heard about the angel and who lost the key of her garage and looked everywhere in her house but couldn't find it. Because she had heard about Archangel Chamuel from the radio show, she decided to ask for his assistance to find her key. She reported later that she received an idea to look somewhere in her house, and she followed through with that guidance and found the key at the exact place where she had been guided to look. She was agreeably surprised at being able to find her key just after asking for the help of the angel.

Notice here that you will never know exactly how or when Archangel Chamuel will bring the things that you are looking for to you. With the example of the shoes, the shoes came to me weeks later in the way that I was expecting the least by the intermediary of a relative who offered them to me free of charge. In the case of the lady who found her key, she received intuitive guidance to look somewhere and immediately found her key. You will never know how and when the things will come to you, but they will come one way or another. I have learned to trust and often tell myself, "I have asked Archangel Chamuel to help me find this," and then I let go and simply wait to see how the things will come, and amazingly, they always come. I will say to keep asking and relax and trust because what you are looking for will come to you sooner or later in one way or another.

<u>When to call upon Archangel Chamuel</u>

If you are looking for a missing object, no matter what that object is

If you are looking for a missing person or pet

If you are looking for the right place to live

If you are looking for your soulmate

If you are looking for the right business partners, the right job

For anything that you may be searching and looking for in life

<u>Prayer to invoke Archangel Chamuel's help:</u>

Thank you, Archangel Chamuel, for helping me to find what I am looking for in life, which includes: _____ (fill in the blank with what you are looking for such as 'my key,' 'the right place to live,' 'loving relationships,' and so on).

Archangel Uriel

Archangel Uriel's name means 'The Light of God.' He is the angel that helps lightworkers and everyone to bring light and love into the world. He is often seen clairvoyantly holding a lantern. His light is golden yellow, similar to the color of a flame. Actually, when Archangel Uriel is around, you may even feel the warmth of his presence. One of his specialties is to give creative insights, new ideas, or inspirations to people. Therefore, Archangel Uriel is the perfect angel to call on if you desire to receive creative solutions, insights, ideas, and thoughts, and if you need help to manifest something in life. He does bring solutions to situations. He also helps to illuminate situations that seem dark or confusing by bringing clarity and understanding. He also protects people by surrounding them with his golden light, which deflects any negativity.

<u>When to call upon Archangel Uriel</u>

When you need creative insights or ideas

When you need help to manifest your goals in life

If you need clarity and understanding of a situation

If you need ideas and answers

If you need motivation and encouragement to reach your goals

If you find yourself in a situation that seems dark and that needs illumination

<u>Prayer to invoke Archangel Uriel's help:</u>

Dear Archangel Uriel, please give me clear insight and guidance for the next step to take.

The other Archangels are:

Archangel Jophiel	Archangel Haniel
Archangel Zadkiel	Archangel Azrael
Archangel Jeremiel	Archangel Raziel
Archangel Raguel	Archangel Ariel
Archangel Metatron	Archangel Saldaphon

As I stated previously, there are countless numbers of Archangels known and unknown to us. However, the '12 Archangels,' which in fact are 15, are the main archangels assigned to the earth to help and guide humanity. They helped humanity in the past as reported in many spiritual scriptures; they are on Earth helping today, and they will continue to do so throughout eternity.

Ascended Masters

I refer to "Ascended Masters" beings from higher dimensions who have graduated from the earth planes and who now serve to guide the earth on the spiritual plane. There are countless numbers of them, hundreds of thousands of them, probably more. The most known of them are Jesus, the Buddha, Moses, Yogananda, Mother Mary, Saint Francis of Assisi, King David, Saint Augustine, King Salomon, Mother Theresa, Joseph, and so on. Some of the ascended masters with whom I worked are Jesus, Mother Mary, Saint Augustine, Yogananda, Saint Germain, and the Buddha. I would like to add here that there are many people who work with the archangels and the ascended masters and that I am not the only one. Christ Consciousness, whom we know as Jesus, has worked with and talked to countless people throughout history, and He is still doing so today. Like God, He is unlimited and is everywhere. I also would like to add here that many people, especially Christians, are praying and calling on the help of Jesus every day all over the world. If you are a Christian and often pray to Jesus, let me reassure you that He listens to all your prayers and often answers them. Therefore, in a sense, Jesus is talking to you as well.

DR. LILLY KOUTCHO

Jesus (Christ Consciousness)

Jesus is now pure Consciousness, and He is one of the main beings who looked after this planet and who are helping countless people on Earth. He is mostly known for His service and His death on the cross, and now He is at the front of a religion, Christianity. However, Jesus does not help only Christians but also helps all who call upon Him regardless of their religions. He helps many people for healing, comfort, and protection.

In the beginning of my awakening, I started to spend a lot of time praying and meditating. Later, I bought Archangel Michael's App of Doreen Virtue and later the apps of other archangels and Jesus. I realized to my great surprise that the archangels and Jesus started to send me personal messages. I almost felt like they were bypassing the writing and sending me direct messages. Later, I would see them in my dream, in my meditation, and receive messages from them through my intuition. I thought that I was hallucinating and wondered why all these beings were around me and talking to me. However, I enjoyed receiving messages from them, as their guidance was always loving, helpful, and comforting. They started to help me to heal and also healed others. For instance, one day, my daughter, who was three years old at the time, accidentally walked on a piece of glass on the floor, cut the sole of her foot, and was bleeding and crying. Out of panic, I started to pray, laid my hand over her foot, and asked for the help of Jesus and Archangel Raphael, as I knew that they were around me. Before I even finished the prayer, the bleeding stopped, the wound was healed, and the sole of her foot was as if nothing happened. My daughter started to smile and jumped on the floor as if nothing had happened. I checked and rechecked her foot again in disbelief, and yes, the wound had disappeared! Oh my God! I couldn't believe it.

Later, Jesus, along with the archangels and other beings, helped me to heal many people and clients. They not only helped me for healing but for various other things such as helping those who are in need. I remembered one day, in the middle of the night, I was with a friend walking on a road of downtown in Kansas City, Missouri, and was guided to go to some specific direction. As we took that road, I met a homeless who approached me, shaken and frightened, and asked for money. I gave him a few dollar bills and then left. We then stopped by the sandwich restaurant nearby to use their toilet. When I got out of the toilet, I saw the homeless to whom I just gave the money sitting there with a sandwich in his hands. I was touched to tears, as I realized that he was walking all night long and was hungry, with nothing to eat. I learned later that it was Jesus and the archangel team around me who guided me there that night to help that homeless, to give him some money so that he could eat. Of course, the homeless had no idea that as he was wandering, hungry, and lost on the road that night, Jesus and the angels were guiding people that could assist him. This is just one example of the many things that these ascended masters and angels are doing behind the scenes to assist people. I worked with them for healing but also for other things such as helping people in need. I learned later that I have worked before with Jesus and the archangels during my past lifetimes.

<u>When to call Jesus</u>
When you need help for healing

When you need comfort and guidance

When you need help to feel God's presence and to remember your divine nature

When you need protection

When you are going through a challenging situation and need support and strength

<u>Prayer to invoke Jesus' help:</u>

Dear Jesus, thank You for watching over me, my loved ones, and the world.

THE TRUTH ABOUT LIFE

THE BUDDHA

Born in approximately 567 BC, the Buddha (real name: Siddhartha Gautama) is one of the main beings who watch over Earth today. His teachings and service work led to the creation of the religion 'Buddhism' after his death. He taught about compassion, which is a form of empathy coupled with understanding. I will say that compassion is closely related to love, which was the main message of another ascended master: Jesus.

The Buddha taught that "the path to peace is through mental discipline" and I couldn't agree more. It took me decades to realize this truth about life. Indeed, having gone through hardships and traumas in life, I suffered a lot, especially when I think about the past and all the things that happened to me. This suffering lingered until one day one of my angelic beings, with whom I work, told me: "It is not the past or what happened that creates your suffering and pain, but it is the thoughts about the past." It took me a little while to fully understand what my angel meant, but I got it. Indeed, when I decided to stop thinking about past hurt and the past in general and only focus on the present moment, the pain disappeared and I realized that my pain and suffering were created by my own thoughts about what had happened. Undeniably, I was feeding the pain and perpetuating my own suffering by thinking about the past, while in truth, what happened no longer exists, it's in the past, gone, over. I realized that it was only in my mind that I was still reliving past experiences; it is only in my mind (thoughts) that it was still happening, and that is where the suffering and pain comes from. "Thinking" is what creates our suffering on Earth, and this "thinking" is what I sometimes refer to as "the voice of the ego", or simply "the ego."

It is fascinating to realize that one of the main teachings of the Buddha was that freedom and the path to peace in life are through mental discipline, and that was ~ 567 years BC. However, this fundamental truth is needed in our world today more than ever before. Because of this truth, the Buddha taught people how to meditate so that they can calm their minds and cease the suffering created by their thoughts or the ego. Meditation is still one of the main practices of Buddhism today.

One important concept taught by the Buddha that I particularly love is "The Eightfold Path", which consists of eight great practices to live by, they are: right view, right resolve, right speech, right action, right livelihood, right effort, right mindfulness, and right concentration. The "Eightfold Path" was the first teaching of the Buddha that I came across, and I was astonished by the wisdom behind His words. I knew instantly that these words were coming from a heavenly being. Just the "right action" alone encompasses the whole "10 commandments" of Christianity. To fully understand the "Eightfold Path", one has to read and understand what these concepts are. I may expand on and explain the concepts of the "Eightfold Path" taught by the Buddha later in my future books.

Another important teaching of the Buddha is the concept of "incarnation and reincarnation", and it is this concept that mainly differentiates Buddhism from other religions such as Christianity, but at their cores, they all speak about love, compassion, and kindness. However, the truth is that Jesus also taught about reincarnation, but unfortunately, these parts of his teaching were removed by the rulers of the Roman Empire, who were avid of control, believed that such teaching was against the church's view, and were concerned that the population would rebel against their supremacy if they understood and realized that they are immortal beings who reincarnate many times. In truth, a population that truly understands that they reincarnate can become unafraid of death and protest against rules and overthrow governments and kingdoms.

The truth is that all these iconic ascended masters came to teach the same things: God's love, peace, compassion, the truth about life, and the path to freedom. However, after their deaths, various religions were created in their names. Some people still believe that the Buddha is God, a deity that they need to worship, in the same way that some believe Jesus is the only Son of God, while the truth is that we are all children of God, all extensions of God, and that the Buddha, just like Jesus, is just one of the highly evolved beings who came to Earth to guide and serve humanity. Later on, religions were created in their names, and today, people who belong to these different religions fight among themselves over which one is right and which one is wrong. This is ironic because in truth on the nonphysical plane, the heavenly plane where these ascended masters live today, Jesus and the Buddha actually work closely together for the betterment of the world. It's funny when you realize that many on Earth are still divided, argue, and fight among themselves, making statements such as, "No, Jesus and Christianity is the only path to heaven, and Jesus is the way," or, "No, the Buddha is the only one sent by God; He performed miracles; Buddhism is the only true religion," etc., while in truth, Jesus and the Buddha are really good friends in their heavenly plane and work closely together to assist people with healing, comfort, and peace on Earth. It makes me smile when I think about this ironic situation.

I love the being that we call "The Buddha", and He is among the heavenly team with whom I work, and I affectionately refer to Him as the "Beloved Buddha". Born and raised in Christianity, I have to admit that I used to be very closed-minded and used to believe that the only true religion that existed was Christianity as I knew nothing about the Buddha. In fact, I had never heard about ascended masters and only knew and believed what I was taught in church. It was only when I had a profound life-changing out-of-body experience and started to communicate with my guardian angels and heavenly guides, that I discovered to my great astonishment the truth about the being who we know today as the Buddha. The Buddha is indeed a highly evolved being sent by heaven to help humanity and to bring peace on Earth, and He is one of the main ascended masters who guard, protect, comfort, and guide us on Earth today. He has helped me countless times with healing, comfort, and guidance, and I can truly say from my personal experience, I who was born and raised as a Christian, that you really don't have to be a Buddhist to be helped by the Buddha. Just by calling His name, you can be sure that He will hear your call because He, like God, is unlimited and can help countless people at the same time.

The Buddha has a soft and gentle energy and always has great compassion and understanding for our earthly situations and challenges. He is a really powerful being who today assists people with various life-related issues regardless of their religions or beliefs. He is a wonderful being to call on if you are going through challenging times in life or when you need help to quiet your mind or when you are sick and need healing.

<u>When to call on the Buddha:</u>

When you are going through challenges in life and need peace and comfort

When you are confused in life and need to see clearly

When you need healing

When you need help to quiet your mind

THE TRUTH ABOUT LIFE

<u>Prayer to invoke The Buddha's help:</u>

Beloved Buddha, please help me to see the blessings in my life; please help me to let go of pain and suffering and find the path to peace. Please help me to focus my mind on what I am grateful for and give any painful thoughts to You and God as soon as they occur.

DR. LILLY KOUTCHO

Yogananda

Yogananda, known as Paramahansa Yogananda, is an ascended master who was born in India before moving into the United States and who performed great deeds and transformational teaching in the world. He founded the Self-realization Fellowship in 1920, introduced Eastern spiritual wisdom into the Western world, and was acclaimed for his life story book, *Autobiography of a Yogi*. However, Yogananda, like the other ascended masters such as Jesus, is still well alive and is working relentlessly to assist people all over the world, not just in India or in the United States. He helps people who need healing and peace, inner peace and outer peace. Several years ago, I was accidentally introduced to one of his books, *Scientific Healing Affirmations*. It was a small booklet that someone gave to me because I was going through tremendous stress in my life and was living in fear, and the person told me that reading that book would help me. Indeed, reading this little book calmed my mind, ceased my fears, and helped me heal from illnesses. I was carrying it with me at all times and would read it during my lunch breaks at work to find peace and comfort. This book became part of my life, and I read the affirmations daily and ended up memorizing them. Today, I know that it was not an accident at all. I often asked Yogananda for assistance for many things, notably for healing and inner peace, and He is part of the celestial team that surrounds me.

I will reiterate here that Yogananda, just like Jesus and the other ascended masters such as Mother Mary and the Buddha, is unlimited and can help many people at the same time and that you don't have to be a yogi and live in the Eastern world to garner his help and assistance.

THE TRUTH ABOUT LIFE

Just by asking for his assistance and help, he will intervene on your behalf and help you.

Yogananda's messages and book are about healing, how to manifest, and how to reclaim your divine nature and your God-given power. He taught the truth about life; how we reincarnate; how to use your divine power to manifest health, abundance, and peace in life. Discovering his teaching was life-changing for me as I started to apply the principles that he taught. I discovered later, that I was not the only one who was using the teaching of the Yogananda to transform my life and manifest things in life. In fact, Steve Jobs, the pioneer of the microcomputer revolution and the founder of Apple and iPhone, was using the same principles taught by Yogananda. In fact, it was reported that Steve Jobs discovered the book of Yogananda, *Autobiography of a Yogi*, when he was 17 years old and read this book every year since he was 17 years old! Before his death, he wrote and suggested that everyone who assisted at his funeral be given a copy of that book. Even a great inventor such as Steve Jobs knew the truth about life, knew that he was a soul who incarnated into this world for a time only for a purpose, and was applying the spiritual principles taught by Yogananda to transform his own life, to manifest, to create all the things that he invented, and to become who he was during his lifetime. It was only after his death that the truth of the principles that he was applying in his life came out.

<u>When to call on Yogananda:</u>
When you need healing
When you need peace, inner peace
When you need help to quiet your mind
When you are dealing with fear, anxiety, and need help to overcome fear

When you need help to rediscover your divine nature and your connection to God

When you want to manifest abundance in life or create something

Prayer to invoke Yogananda's help:
Beloved Yogananda, please bring perfect peace and understanding to my mind, my situation, and my life.

The new generation of ascended masters:

When people think about Ascended Masters, they often think about Jesus, the Buddha, and those well-known ascended masters who came to Earth thousands of years ago to serve, teach, and guide humanity. Many of us don't often think about the new generation of ascended masters, and to be honest, only a few people know about this new generation of ascended masters. However, the truth is that ascended masters also comprise these souls who, after finishing all their learning on Earth, graduated (ascended) and no longer need to come back to the Earth school, yet volunteer to come back to serve by carrying a specific mission that will help to propel the Earth in its evolution. In truth, there has always been ascended masters coming to Earth, and a brand-new generation of ascended masters are coming in these modern times. One of this new generation of ascended masters is Mother Teresa.

THE TRUTH ABOUT LIFE

Mother Teresa (Saint Teresa of Calcutta)

Mary Teresa Bojaxhiu, commonly known as Mother Teresa, was born in Skopje, North Macedonia in 1910, and is mostly known for her charitable work, her care for the poor, and her kindness. Mother Teresa was a catholic nun who decided against all odds, obstacles, adversities, and challenges to care for the poorest of the poor among us, the most vulnerable in society, the unwanted, the lonely, the abandoned, the neglected, the ones labeled impure, the casted ones, and those called unworthy. Many of us won't even dare to walk close to these people because we are too afraid of them, and many will not even look at them; yet Mother Teresa went where the homeless, people dying from HIV/AID, the hungry, and the abandoned people on the road lived, held them in her arms, washed their feet, treated their diseases, and fed them.

Most of us know Mother Teresa as the now canonized catholic nun who conducted charitable work around the world. In truth, Mother Teresa is a graduated soul (ascended master) who came to Earth to teach us what love truly means. She came to teach us about humanity, compassion, mercy, and kindness. She was a resilient soul who decided to live among the most vulnerable of society by entering their worlds: she washed the sores of children, cared for sick old people who were abandoned on the streets, nursed those dying from tuberculosis, and rescued the homeless from the streets. Yes, Mother Teresa was resilient in her mission, and her legacy still lives today to inspire us and show us what we are all capable of: "Love".

The strength, determination, resilience, and kindheartedness of Mother Teresa reminds me of a firefighter who decided to rescue a little girl trapped in a house on fire, and even though he knew that it would be challenging to rescue the little girl, he decided anyway to jump in to save her by saying: "God, I am going in that burning house, please back me up; please help me to come back alive with that little girl because I cannot stand here and watch her burn alive for fear for my own life. So, I am taking a leap of faith; I am going into the burning house, and all my hope is in You" and God heard the firefighter's cry for help and backed him up: the firefighter came back from that burning house with the little girl alive and unharmed.

Yes, God heard the heart of Mother Teresa and backed her up, and she went among the unwanted, the cast, and she came back rescuing them. Indeed, Mother Teresa successfully completed her mission of service to humanity by bringing back to her home, "Home of the Pure Heart", the neglected and those abandoned by society, and she treated them like angels. The hungry were fed; the sick were treated medically; the dying were given the opportunity to die in dignity and love according to their religious beliefs: Muslims were read the Quran, and Christians were blessed with the holy unction because Mother Teresa wanted to make sure that these people who lived like animals died like angels. And God backed her up! Mother Teresa is a hero; she is a hero of humanity; she is a hero of demonstration of what love truly means. She is still today a living demonstration of inspiration for all of us. She is an ascended master.

THE TRUTH ABOUT LIFE

Dr. Martin Luther King Jr.

Dr. Martin Luther King Jr. was an American activist, an influential spokesperson, and a leader of the civil right movement in the United States. He was mostly known for his powerful and outstanding speech, "I have a dream", a speech that was a call for economic and civil justice and for an end of racism and segregation in the United States. He was assassinated the following day by a gunshot for his beliefs and for advocating for justice and equality.

What we don't know is that Dr. Martin Luther King Jr. is an ascended master whose mission was to bring social justice and dignity to a place in the world (the United States of America) where there had been centuries of segregation, discrimination, and humiliations of those souls, of those children of God who were labeled "Negroes". So, heaven sent an ascended master to trumpet and teach the fundamental truth that "all men are created equal." This was the mission of Dr. Martin Luther King Jr.; that was his assignment. He came to proclaim the good news to the poor, to bind up the broken-hearted, to proclaim freedom to the captives of segregation, to release his people from the darkness of discrimination, to set these prisoners labelled "negroes" free from their pit of despair, to lighten their burdens, and to give them hope and a new future.

Yes, that was the mission of Dr. Martin Luther King Jr., and he carried his mission with confidence, courage, and an extraordinary fearlessness. He was unstoppable in his mission despite the challenges, threats, and even imprisonments. For a man who was seeking justice, peace, and freedom for all, he was jailed more than 20 times, stabbed once, his house was firebombed, and he suffered from multiple attacks and constant threats, yet he continued his mission for social change, freedom, and justice. He put his life on the line so that his brothers and sisters could be free one day. I will say that even if Dr. Martin Luther King Jr. was not nailed to the cross, nevertheless, he laid down his life for others, for freedom, for justice, for equality.

In truth, the message that Dr. Martin Luther King Jr. was advocating and teaching during his life was the "Oneness principles". Indeed, the main message of Dr. King was that all men are created equal regardless of their race, skin color, gender, or social background: this is the basic core of the Oneness Principle 3: "We are unique and yet equal." In the 'Letter from Birmingham Jail', which was a letter written by Dr. Martin Luther King Jr. while he was imprisoned in Birmingham jail for coordinating pacific nonviolence marches and sit-ins against racism and social segregation in Birmingham, he stated, "Injustice anywhere is a threat to justice everywhere," which is simply an exemplification of the Oneness Principle 5: "We Are All Connected".

Below is a summary of the "Oneness principles", which were explained in the first chapter of this book:

Oneness Principle 1: We Are All One

Oneness Principle 2: We Are All the Same

Oneness Principle 3: We Are Unique and yet Equal

Oneness Principle 4: We Have the Same Potentiality

Oneness Principle 5: We Are All Connected

Oneness Principle 6: We Are All Part of a Greater Life Even Though We Have Our Own Individualized and Complex Lives

Please go back to Chapter 1 of this book to reread the details of the "Oneness Principles" if needed.

THE TRUTH ABOUT LIFE

Some may be surprised, argue, or doubt if Dr. Martin Luther King Jr. is truly an ascended master because when we think about ascended masters, we often think about big religious figures such as Jesus or the Buddha who came to accomplish spiritual works and who today are in the front line of gigantic religions. Some may be surprised that an activist and leader of the civil right movement is listed among ascended masters such as Jesus or the Buddha. However, I want to remind us here that the Buddha was also considered as an outsider and a preacher of freedom when he was alive as He rebelled against the rules of his society and left the royal palace where he was born, leaving behind the luxurious lifestyle of His royal family to live an eccentric life of deprivation and teaching His message of peace and compassion. Furthermore, Jesus, too, was considered as a "rebel" by the religious and political leaders of His time because of His refusal to follow the pre-established rules of society and for his fearlessness in speaking the truth so much that they captured Him and nailed Him on the cross for His "disobedience".

Yes, Dr. Martin Luther King Jr. is one of them: He is one of the ascended masters.

Just like the apostle Paul went from city to city in the Graeco-Roman world preaching the gospel of love, Dr. King went from city to city across the United States preaching the gospel of freedom and justice.

Just like the apostle Paul was imprisoned for preaching the gospel of love, Dr. King was imprisoned more than 20 times for preaching the gospel of nonviolence and non-segregation.

Just like the apostle Paul was writing letters while he was in prison, which became what we know today as the Epistles of Paul or the books of Ephesians, Romans… etc., Dr. King was writing in prison the 'Letter from Birmingham Jail', with the hope that his message of freedom and emancipation for the black community will be heard one day.

Just like Jesus was crucified for preaching the message of love and forgiveness, Dr. King was assassinated by a gunshot for preaching the message of freedom and peace.

Yes, Dr. Martin Luther King Jr. came to proclaim the gospel of freedom, the gospel of equality and social justice. He was a man of God; He is an ascended master.

For those who will still question why or how a black man can be an ascended master, I will say to them: you have missed the main message of Dr. Martin Luther King Jr: "All men are created equal"; there is no racial segregation in heaven.

Below is an excerpt from the **"I Have a Dream" Speech** of Dr. Martin Luther King, Jr., delivered at the Lincoln Memorial, in Washington, D.C.

And if America is to be a great nation, this must become true.

> *And so let freedom ring from the prodigious hilltops of New Hampshire.*
> *Let freedom ring from the mighty mountains of New York.*
> *Let freedom ring from the heightening Alleghenies of Pennsylvania.*
> *Let freedom ring from the snow-capped Rockies of Colorado.*
> *Let freedom ring from the curvaceous slopes of California.*

> *But not only that:*
> *Let freedom ring from Stone Mountain of Georgia.*
> *Let freedom ring from Lookout Mountain of Tennessee.*
> *Let freedom ring from every hill and molehill of Mississippi.*
> *From every mountainside, let freedom ring.*

And when this happens, when we allow freedom ring, when we let it ring from every village and every hamlet, from every state and every city, we will be able to speed up that day when all of God's children, black men and white men, Jews and Gentiles, Protestants and Catholics, will be able to join hands and sing in the words of the old Negro spiritual:

Free at last! Free at last!
Thank God Almighty, we are free at last!

Let freedom rain

A prayer for humanity: in appreciation and gratitude for Dr. Martin Luther King Jr.

I want to thank Dr. Martin Luther King Jr. for his service and sacrifice because I would not be where I am today, sitting here in the comfort of my home in Tulsa, Oklahoma, in the United States of America, writing these words if he hadn't done what he did. Indeed, I was only a black girl born into a poor family in Africa, who moved to the United States of America with the hope of living the American dream.

And yes, I lived my American dream, and I am still living it today. Indeed, I was able to attend the pharmacy school of my dreams down there in a city of Kansas, a white folk city, with more than 150 white students in the classroom and only 2 black students: I was one of those 2 black students.

Yes, the ring of freedom is ringing today from every state and every city of the United States, thanks to the work of Dr. Martin Luther King Jr. So, today, I want to say, "thank you".

However, am I satisfied? No, I will never be satisfied as long as there are still children, women, and races that are still victims today (in the year 2019) of marginalization, atrocities, abuse, abandonment, slavery, and rape right here, right now on our planet Earth. No, I am not satisfied; even though the ring of freedom is softly ringing today, I believe that we need more than a ring: we need "a rain of freedom". The rain of freedom must pour down on Earth like mighty water to cleanse and purify the wounds that humanity still has today. So, I want to say:

Let freedom rain

Let freedom rain down there on these small villages of Africa where mothers are crying because their children are hungry and have nothing to eat. Let freedom rain.

Let freedom rain down here in the United States where there are still people today who are working 8 hours a day, 5 days a week, 50 weeks a year, only to be called the "working poor" and are unable to live a life of dignity and are enslaved by the chains of poverty. Let freedom rain.

Let freedom rain down there in the Middle East where women are still being forced into marriage and crippled by the manacles of discrimination of the male gender because of the belief that women are "less than" men. Let freedom rain.

Let freedom rain down there in India where a whole group of people is seared in the flame of injustice, rejection, and humiliation because of a system called "caste system", which deems them impure, uncleaned, and cast. Let freedom rain.

Let freedom rain down there in Benin, Ghana, Cape town, Abidjan, where there are little children of 7, 9, and 12 years old living on the roads day and night, begging for food because we are unwilling and incapable of coming up with a plan to house and feed these vulnerable children of God. Let freedom rain.

Let freedom rain down there in the Middle East where our brothers and sisters are being killed and burned alive because they are gay, lesbian, or transgender. Let freedom rain.

Let freedom rain there down in Lomé, West in Africa, where smart A+

students are graduating from university only to become these struggling motorcycle taxi drivers called "Zemidjan". Let freedom rain.

Let freedom rain down there in those prisons in the United States, Europe, Africa, and Asia where there are still God's children locked like dogs in a cage, living like animals day and night, year after year, decades after decades until death releases them from their pain and agony. Let freedom rain.

Let freedom rain down there on the streets of our capital city, Washington, D.C., where there is still today what they call "child prostitution", where 12-year-old girls are walking at nights at 2 a.m. on these streets, forced to prostitution by those called "pimps". Let freedom rain.

Let freedom rain down there on these regions where people are still being killed today because of their religious beliefs. Let freedom rain.

Let freedom rain down there on Darfur, in Sudan, Northeast Africa, where till today families are being killed, women are being raped, and their children kidnapped and sold as slaves just because others who live in this country have labeled them "the inferior ethnicity". Let freedom rain.

Let freedom rain on all of these countries that are still spending today billions of dollars to develop atomic bombs with the insane hope that, one day, they may drop these bombs on their fellow brothers and sisters to win a war. Let freedom rain.

Let freedom rain on every blue-collar county of America.
Let freedom rain on every cold city in Europe.
Let freedom rain on every abandoned village in Africa.
Let freedom rain on every forgotten suffering home in Asia.
Let freedom rain from Tulsa Hill of Oklahoma to Zangéra in Lomé!

Let freedom rain!

DR. LILLY KOUTCHO

Saint Francis of Assisi

Saint Francis of Assisi is one of the world's most popular and beloved Saint. In fact, I have traveled to many countries and cities in Europe, North America, South America, and Africa and have encountered again and again many hospitals, churches, and streets named after Saint Francis of Assisi: "Saint Francis hospital", "Saint Francis church", "Franciscan", "Saint Francis street", etc.

Saint Francis of Assisi is indeed very popular among the Catholic congregation as well as the non-Catholics as testified by the fact that many people around the world continue to name their children "Frank, Francis, Francisco, Franco, Francisca…" etc., in honor of the legacy of Saint Francis on Earth. His popularity around the world is undeniable, and even our actual pope, Pope Francis, chose the name "Francis" in honor of the work and service of Saint Francis of Assisi.

Saint Francis of Assisi was born Giovanni di Pietro di Bernardone in Italy around 1181/1182 to a wealthy Italian family. He later gave up his wealthy lifestyle to devote his life to serve the poor and the sick. He became a catholic friar and a preacher who stood by his compassion, his service to the poor, and his love for animals.

Saint Francis is an ascended master who came to perform great deeds of service to humanity and to teach us to treat everybody with respect, dignity, and love regardless of their socioeconomic status or physical appearance. He was known to care for underprivileged people. He performed his teaching by putting into action what he was preaching—by helping the needy on the streets and those in despair.

Saint Francis of Assisi also rejoiced in the value and beauty that animals bring to Earth and asked everyone to love and praise all God's creatures, including animals. Indeed, Saint Francis of Assisi was known for his love for animals, and today, he is recognized as the Saint Patron of animals and ecology. It is said that Saint Francis of Assisi was so loving and emanated such peace that birds landed still in His arms. Because of his love for animals, he is often depicted surrounded by birds or other animals in many paintings and artworks.

I was born into a Catholic family, was baptized and went to a church named "Franciscan" for many years, and later worked as a pharmacist at "Saint Francis hospital", but ironically, I didn't know who Saint Francis truly was. I came across the work of Saint Francis one day when I was reading a book of Dr. Wayne Dyer, the internationally renowned bestselling author, speaker, and pioneer of self-development. In his book, Dr. Wayne Dyer stated that the world's popular "Saint Francis Prayer", also known as "The Peace Prayer", has transformed and changed his life. I had never heard about this prayer before, so I was curious and went to the web to find out what it was about. The words of this prayer deeply touched me, and I decided to pray "The Peace Prayer" regularly. This prayer started to transform my life too and has inspired me to become a more compassionate and loving person. Below is the "Saint Francis Prayer" or "The Peace Prayer":

> *Lord make me an instrument of your peace*
> *Where there is hatred let me sow love*
> *Where there is injury, pardon*
> *Where there is doubt, faith*
> *Where there is despair, hope*
> *Where there is darkness, light*
> *And where there is sadness, joy*

O divine master grant that I may
not so much seek to be consoled as to console
to be understood as to understand
To be loved as to love
For it is in giving that we receive
it is in pardoning that we are pardoned
And it's in dying that we are born to eternal life
Amen.

In truth, Saint Francis of Assisi is an ascended master who came to serve humanity. Even though he is mostly known in the catholic population, Saint Francis is here for everyone and is here to help Catholics and non-Catholics alike, believers and non-believers alike. He is an ascended master and one of the main beings who are guiding and protecting our planet Earth. I believe that Dr. Wayne Dyer, who was an internationally known spiritual teacher and whose work has transformed and changed the lives of millions of people in this world, was working closely with Saint Francis of Assisi on a spiritual plane and has been guided by Saint Francis of Assisi throughout his life. I believe that just like Steve Jobs, the founder and inventor of Apple, was guided by the ascended master, Yogananda, to download and channel technologic information that has revolutionized the world, Dr. Wayne Dyer was guided by Saint Francis of Assisi to download and channel information and materials that has changed the lives of millions of people around the world.

Like the other ascended masters, Saint Francis is still here today, helping all who call on his help, despite their religious beliefs. As the Saint Patron of animal and ecology, Saint Francis of Assisi's message to the Earth was to respect and treat the animals that are in our world with love because they are all God's creatures.

I believe that we, human beings, habitants of Earth, still have a lot to learn and grow when it comes to how we treat animals and our environment. We sometimes act like little toddlers, and we are destroying and wrecking the very home we live in: Mother Earth. Some greedy people and businesses are depleting or destroying Earth's natural resources to make money. Some, by pure ignorance, are still throwing plastic bags, soda bottles, cups, and others items on the ground, and these waste and debris are later found in our oceans, polluting the sea and killing the marine animals and fishes.

This childlike behavior that many of us still have on this planet comes from total ignorance of the "Oneness principles". In truth, the "Oneness principles" do not apply only to human beings but to all God's creatures, including animals and our planet, Mother Earth, that houses and feeds us. Any destructive or uncaring behavior toward animals or Mother Earth will come back to us and slap us in the face like a boomerang: this is the basis of the "Oneness Principle 5", which states that anyone who has any destructive behavior toward anybody or anything, whether humans, animals, or the planet, is hurting himself or herself because we are all connected. Indeed, these wastes, plastic bags, and chemicals that we throw on the ground are later found in the sea that we bathe ourselves in, in the water that we drink daily, in the grass that feeds the chickens which we then consume later, in the fishes that we eat for lunch and consequently ends up causing diseases and allergies in our bodies later on. That is why we must care for our environment and the animals—because we are all connected. We are all part of a Greater Life: One Life.

Unfortunately, there are still today sports/shows such as bullfighting in many areas of our world like Portugal, Spain, Mexico, etc., where people gather to watch these animals get treated poorly as entertainment. Bullfighting is a brutal sport that consists of baiting, provoking, or killing a bull as a public spectacle in an outdoor arena. As shocking as it can seem, we humans are still at a point of our evolution where we find these brutal sports—in which animals are treated cruelly and violently—normal, entertaining, and even pay money and gather to watch and enjoy them. Furthermore, bullfighting is not only dangerous to the bulls involved but also to the bullfighters or riders, as they put their lives at risk as well.

A couple of years ago, I was on vacation in a city of Costa Rica, and it was their annual city fiesta. I decided to go and check out this fiesta that everyone was talking about. Well, it was a bullfighting contest. As I sat there watching, I was saddened by the way the bulls were treated and shocked by the way they are provoked to anger for the purpose of entertaining people in the arena. I was also scared and concerned about the bullfighters and so shocked that I ended up leaving the arena to return to my hotel room.

Thank God we are no longer in the time of gladiators of the Roman Empire when human beings confronted each other or wild animals in violent combats, fighting to death in public arenas where the audience would applaud and laugh for entertainment. Thank God we have moved past that stage, but today, in 2019, we still find bullfighting entertaining and an enjoyable fiesta of the year. There are still people on this planet who yell at their dogs, beat up their dogs, or even torture their dogs to death. We can all honestly agree that we have a far long way to go in our evolution and understanding.

Some may think that the fate and the poor treatments that some animals are victims of on this planet is not heaven's business and that heaven or God (call it whatever you want) only cares about "these religious affairs" such as "love each other", "pay your tithe", "memorize your Quran", "memorize your Bible verses", "go to church every Sunday and sing your 'Alleluia, Alleluia'", or "go to Mecca every year and sing 'Allah hu Akbar, Allah hu Akbar'", but let me assure you that you are mistaken because heaven does care about these animals. In fact, heaven cares so much about these pure creatures of God that they sent us the ascended master that we now know as Saint Francis of Assisi to teach us how to love and treat animals with kindness and respect.

Currently, it is not just the animals that we are mistreating by our ignorance and childlike behaviors on Earth but also our space and the welfare of the surrounding nearby planets, and by doing so, we are disturbing the homeostasis of the whole Universe. Homeostasis is defined as the tendency toward a state of equilibrium between interdependent elements. Thus, a disturbance or imbalance of one element among a group of elements that are related (interdependent) will disturb the equilibrium and the wellbeing of all of the surrounding elements, therefore, disturbing the stability of the whole.

As a biologist and a pharmacist, I know how important homeostasis is for the general wellbeing of the body. Similarly, all the galaxies and the planets that constitute the Universe also need to maintain a state of homeostasis for the well-functioning and harmonization of the Universe as a whole.

However, unfortunately, many human beings still believe today that we are the only species that exists in the Universe or simply, by pure ignorance, they don't care about the wellbeing of our planet Earth and its surroundings and are throwing satellites, artificially created objects, defunct man-made objects, and God-knows-what-else into the space, and these objects become what people call "space junk", "space waste", or "space garbage" that orbit around the Earth, polluting the space and endangering the wellbeing of planet Earth as well as the welfare of the other nearby planets, consequently disturbing the homeostasis of the whole Universe. In fact, currently, there is an estimate of over 170 million pieces of space debris, manmade satellites, and other junk thrown into space by humans that are now orbiting around the Earth. Yes, an estimate of over 170 million pieces of space junk launched into space by us, human beings!

I didn't even know what "space junk" was and never heard about it until a few years ago when, on a nice summer night, I decided to go outside, hang out on my backyard, and look at the night sky to contemplate and admire the beauty of the stars and our Universe. I saw something in the sky that I believed to be a star, but I was told, "No, this is not a star but a satellite." I saw many other objects in the sky looking like stars, but again, I was told that they were either satellites or "space junk", and I was taught how to differentiate between these manmade satellites and other artificial man-created space junk from the true stars. I was stunned by the number of satellites and other space garbage in the sky. I couldn't believe it.

The sad thing is that I don't think we human beings or our governments are planning on going back to space to clean up the mess that we are creating with our so-called new space technologies. No, we simply throw them out there in space, creating these "space junk", just like some of us after drinking their coffee or soda simply throw the plastic cups and bottles on the ground and move on with their daily affairs without caring about where this plastic garbage will end up. However, "space garbage" or "space junk" is not like simple plastic bags, cups, and bottles; they are spatial objects that will remain in space for centuries or even millennia and keep polluting and destroying the welfare of the whole Universe.

In truth, we human beings often act like little toddlers in the Universe, and not only are we hurting ourselves and killing each other every day, but we are also destroying the very planet that was lent to us so that we can come and learn on. Moreover, we are destroying the beauty of the Universe by throwing satellites and junk into space. This childlike behavior that we often display reminds me of a 4-year-old toddler who likes to play with scissors—cutting out her clothing, the sofa of their living room—and not knowing what she is doing or understanding the impact and consequences of her actions.

Some may think or say: "Oh, I don't care about the Earth, the environment, space, and all these ecologist talks, blah, blah, blah. The fate of the Earth is not my problem. I have other things to care about. I am not going to be concerned about something that may happen in hundreds of years from now because I am not going to be here on Earth by then; I will be dead by then, so I don't care," but guess what? You may come back to this very same planet Earth to continue your learning. Therefore, even if you think it is not your problem today, it will become your problem then—when you come back for your next reincarnation. So, it is wiser to start taking small action steps today to preserve the life of animals and to protect our environment, our planet, and space.

Loving and caring for all God's creatures, including animals and protecting the environment, was some of the missions of Saint Francis of Assisi. For those who love animals and who want to volunteer and rescue animals or those who desire to protect our planet Earth, Saint Francis of Assisi is a wonderful being to call on. Yes, Saint Francis of Assisi was sent from heaven; he is an ascended master.

DR. LILLY KOUTCHO

Mahatma Gandhi

Mahatma Gandhi is mostly known as an Indian nationalist, activist, and an adherent of nonviolence, whose effort led to the independence of India from British colonialism. Today, he is recognized as an international symbol of peace.

Mahatma Gandhi was born in India in 1869 and was later trained as a lawyer and moved to South Africa where he worked as an attorney. In South Africa, he faced social discrimination, as the Indian community did not have the same rights as the white population, and was discriminated against in many ways. Shocked by the way his people, the Indian communities, were treated, Gandhi decided to mobilize the Indians of South Africa to oppose prejudice and social injustice using pacific and nonviolent approaches. He later moved back to India where he continued on with his passion as an advocate for social justice by opposing the colonialism rules of the British government.

In truth, Mahatma Gandhi was a graduated soul (an ascended master) who came to Earth to conduct great deeds of peace, unity, and harmony. We can all agree that he had accomplished his mission, as he is still cited today as one of the most influential people who have revolutionized and shaped the history of the Earth. Indeed, Gandhi's approach of nonviolence has changed the face of the Earth for good and has inspired the civil rights movement of the United Stated as well Nelson Mandela's leadership that has led to the dismantlement of apartheid in South Africa. He is still inspiring people all around the world today.

What made Mahatma Gandhi stand out is his unique approach of using nonviolent techniques to resolve political conflicts and bring about social change and justice. Never before has there been such a thing in the history of humanity as wanting to gain independence from a colony without using bullets, killing, fighting, or any violence. Yet, Gandhi believed that it was possible to gain results, to stop discrimination, and to bring about social justice and peace without violence.

I personally like the following quote of Gandhi: "An eye for an eye only ends up making the whole world blind." Can you feel the divine truth behind these words? Indeed, if we have to pay back offense for offense, there will be no end to the madness, and this will only lead to endless fights and wars, which is where humanity is at today: nations fight nations and again and again. However, in truth, forgiveness is the solution that will renew humanity and heal the pain that humanity still holds deep inside. The suffering, the fights, and the wars will only cease when one party or one nation decides to forgive and let go of the offense. We often think that when we forgive, we are going to lose or that forgiving is "being weak", but Gandhi stated: "The weak can never forgive. Forgiveness is the attribute of the strong."

One of the teachings of Gandhi is the resolution of conflicts with spiritual tools of peaceful approaches and forgiveness. Indeed, we have been taught to fight against each other, to hold grudges, but the path to peace is the complete opposite of what we have been told. It is evident that we human beings are on a lost path and we don't know the truth about life as we still believe in the use of weapons and killing as solutions to resolve our disagreements as testified by the many wars and atrocities that have happened on this planet and that are still happening today.

When I graduated from high school, I left Togo, my country of origin, and headed to France to pursue university studies. I lived in France for 8 years, and there, when I was in college studying Neuroscience, I used to work as a home health aide for the elderly on weekends and nights in order to pay for my school tuitions and needs, and I had the opportunity to take care of a few elderly who were alive during War World II and they told me their stories. I was touched by their stories and could feel the pain and the suffering of what they had been through in their voices. I knew they had been severely marked by the events of this war. However, they have all told me that even though it was hard, it was behind them now; they have forgiven, and it was in the past. From my discussions with these lovely elderly people who have faced the atrocities of World II and from living in France for several years, I know for sure that some cities and villages of France still carry the memory of the atrocities of the Nazi massacre of World War II. However, I believe that the people of these two countries (France and Germany) have decided to forgive each other, to opt for reconciliation, to move on, and today, these two nations live in peace, despite the atrocities that have happened between them.

On the other hand, I was just a little girl living in Africa when the genocide of Rwanda happened, but I remember my uncle, who was passionate about history, saying that the people of this region in Africa had had other similar genocides between the two ethnicities involved (the Hutus and the Tutsi) earlier in history and that this was not the first time. Indeed, there had been previous genocides between these two ethnicities, notably the genocide of Burundians Hutus around 1972 by a Tutsi-dominated army. It was said there had been an atmosphere of unforgiveness and hatred born from earlier genocides when the Hutus were killed, which built up until, one day, it exploded, and the Hutus decided to get revenge, which led to the genocide of Rwanda in 1994, with a bitter aftermath of between 500,000 and 1,000,000 deaths. This horrible Rwandan genocide has shown us all how far unforgiveness, hatred, segregation against one ethnicity or one race can go when we, human beings, lose touch with the reality of who we are and with the truth. There have been so many wars, genocides, and suffering on this planet, such as the genocide of Palestine, Darfur, Rwanda, Cambodia, Bosnia, Yazidis, Guatemala, etc., and there are still wars and killings going on here on this planet at this very moment. However, the ultimate truth is that forgiveness and reconciliation is the only way of salvation for humanity. This does not imply only forgiveness between nations or races but also forgiveness between families, between siblings, between husbands and wives, between coworkers; forgiveness for everybody and everyone. Yes, forgiveness is the remedy that will heal humanity, and that was one of the teachings and messages of Mahatma Gandhi.

The tiger analogy

An analogy to illustrate why we must forgive

Let us suppose, for instance, that you have been hurt by a wild tiger. Will you carry resentment toward that tiger and go about your life carrying anger toward the tiger? Will you say, "How could this wild tiger do that to me? I cannot believe what the tiger did! How awful that tiger is?" etc.

Here is the truth: a tiger is a tiger, and a tiger will do what savage animals naturally do. A tiger is not a stuffed toy or a puppy dog that you bring home and play with: it is a wild animal. Therefore, if you come close to a tiger, well, it may hurt you, scratch you, or even eat your leg. A tiger is just a savage animal, and it will act according to its wild nature and its awareness. Carrying resentment toward the tiger is useless and will only hurt "you", not the tiger. In fact, the tiger will simply continue with its wild life without even remembering that it had hurt somebody. That is why you must forgive the tiger because the tiger does not know any better; the tiger does not know what it is doing. Similarly, carrying resentment or remorse toward an offender is useless and will only hurt you, no matter the severity of the offense because the offenders don't know better. These people who hurt others are just lost souls who are suffering themselves, who have been hurt themselves, and who have lost touch with the truth; therefore, they don't know what they are doing. "Then Jesus said, 'Father, forgive them, for they do not know what they are doing'" (Luke 23:34).

I want to mention here that my intention is not to compare human beings' behaviors to tigers' behaviors, even though sometimes we human beings act like tigers toward each other.

We still have so much to learn about forgiveness, and in truth, there are so many misconceptions about forgiveness, and I believe that many of us do not fully understand what forgiveness truly is and its importance in healing, in manifesting abundance, and in our spiritual evolution, which are the very reasons why we are here on Earth.

Eight big misconceptions about forgiveness:

Forgiveness misconception #1: Forgiveness means forgetting the offense

Many people think that forgiving means forgetting the offense, and sometimes they struggle to pretend that nothing has happened: this is not forgiveness but "denial". Forgiveness does not mean forgetting what happened. The truth is that what happened in the past has happened. The act of forgiveness is finding a way to transmute and transcend the event so that you can live a happy life by not carrying past hurt. "Forgiveness does not mean dwelling upon the wounds. It means looking at the world through unwounded eyes." Forgiveness is admitting that the hurt has happened and yet finding a way of healing your mind, healing your heart, and getting to a point where the thoughts, the pain, and the past do not arise anymore in your mind to create pain and suffering.

Forgiveness misconception #2: If you forgive, it means that you condone the behavior

Forgiveness does not mean condoning the offense. Forgiveness means acknowledging that the act of the other person is "not okay", yet because you love yourself, deciding to forgive anyway in order to cleanse and free yourself from the bondage of past pains.

Forgiveness misconception #3: Saying "I forgive you" to the person means that you have forgiven the person, and it is enough

Forgiveness does not mean saying "I forgive you". Today, we live in a society where faking and trying to be nice has become the norm. Sometimes, we are so concerned by outside appearances that we say to others that we have forgiven them, only to ruminate about the offenses later in our minds and hearts. However, what matters the most is our true feelings. Forgiveness is a deep inner work that requires honesty, sincerity, and a strong willingness to forgive. However, even though saying "I forgive you" does not mean that you have necessarily forgiven the person, saying it does have a positive effect and may prevent further fueling the situation while working privately on yourself to forgive that person.

Forgiveness misconception #4: Forgiveness means trying to be nice

I often hear people say: "I have to be the bigger person, so I have to forgive them". Therefore, they fake their forgiveness in order to appear like "a nice person". Sometimes, while doing so, they don't truly forgive and end up with repressed emotions, anger, and hurt feelings that remain unaddressed. The truth is that any unaddressed feeling or emotional pain will bubble up later in another form and will follow you until they are addressed, released, and healed. Forgiveness truly means "releasing the persons from your emotional grasp and having no negative feeling attached to them anymore". It is truly doing the inner work to become completely at peace with the situation or the persons involved.

Forgiveness misconception #5: Forgiveness means remaining in the situation or staying in relationships with those who have hurt you

You can forgive a person but at the same time not having a relationship with that person anymore, as long as your decision to distance yourself from that person was not made out of anger or resentment. I often see many abused women who remain in the abusive relationships because they falsely believe that forgiving their abusers means staying in a relationship with their abusers again because they think: "After all, God says to forgive and love again". The truth is that God does not want anyone to suffer or to be abused. This is not God's will for anyone, and any type of abuse is unacceptable.

Forgiveness misconception #6: Forgiveness is a "one-time" act

Many falsely believe that forgiveness is a "one-time" act and that when they have forgiven once, it is done once and for all. The truth is that forgiveness is an ongoing practice; it is something that you will encounter again and again throughout your life. Sometimes, you will have to forgive the same person 400 times. In truth, forgiveness is a day-to-day practice. "Then Peter came to Jesus and asked, 'Lord, how many times shall I forgive my brother who sins against me? Up to seven times?' Jesus replied, 'I tell you, not just seven times, but seventy-seven times!'" (Matthew 18:21). This number "seventy-seven times" is simply a metaphor to illustrate that we may need to forgive an infinite number of times as long as forgiveness is needed.

Forgiveness misconception #7: The person that needs to be forgiven has to deserve it

Oftentimes, we refuse to forgive others, thinking that they do not deserve it. This is especially true when the offenders are not repentant, are not regretful, or refuse to apologize or recognize their faults. The truth is that some people will never recognize their wrongdoing and will never apologize; they may even stick to their positions and state that they have done nothing wrong, even though they know in their hearts that their actions were not okay because they are too proud of themselves or just because that is simply who they are: "lost souls with a tiger's attitude". However, that is okay because you don't need their apologies or repentance in order to forgive them. You simply have to accept that you may never get an apology and be at peace with that. This does not mean that these people do not deserve to be forgiven; it simply means that these people are just who they are, with the level of awareness and understanding that they have on this time and space of their evolution. Forgiveness is a gift that you give to yourself, and it has nothing to do with them, their actions, or their willingness to apologize. It is for your own peace of mind, your own freedom, and for your own relief from suffering.

Forgiveness misconception #8: Having an entitlement consciousness

The entitlement consciousness is when we believe that people owe us something, and when we have these beliefs, we suffer, and we struggle to forgive. Indeed, sometimes, we believe that people owe us certain treatments such as loyalty, respect, supportiveness, money, love, or even an apology. Sometimes, we get mad when someone we know passes by us and willingly refuses to say "hello" to us. I used to be frustrated with these unkind behaviors of others until, one day, I realized and accepted that "they don't owe me a 'hello'". What a relief! It is indeed a relief to understand and accept this truth of life.

The truth is that no one owes us anything, not even an apology, love, or a 'hello'. For instance, I used to struggle for many years with the fact that my mother does not love me as she loves my other sister; after all, isn't it a mother's responsibility to love all her children? And I had developed the feeling of entitlement toward my mother until one day the truth hit me: she does not owe me anything, not even love. It was a bitter pill for me to swallow, but it was the truth. It was as soon as I accepted that truth that I was really able to begin my forgiveness journey and heal.

I remember another incident, when a team with whom I worked, whom I loved, deeply trusted, and considered to be my family dumped me out of the team (fired me) in a crucial moment of my life, fully knowing that I had no financial support to survive. I was shocked, hurt, disappointed, and saddened, and I had a hard time forgiving them because, once again, I had developed the "entitlement consciousness" that they owe me loyalty and friendship because, alas, I wholeheartedly believed in their integrity, trusted them, and considered them as the family that I never had. Therefore, it was very hard for me to forgive them, but I had to accept the truth that they owe me nothing, not even loyalty, consideration, or kindness.

Yes, no one owes you anything because people have the freedom or "free will" to misbehave and to be unkind: this is the truth. Even though there are consequences in life (the law of cause and effect), people still have the right to use their free will to do whatever they want. Accepting and understanding this truth of life is healing for the mind and the soul.

What does forgiveness truly mean?

Forgiveness does not mean "what they did is OK", but it means "I am no longer willing to hurt myself by harboring toxic anger".

Forgiveness means detoxifying your body and mind of toxic anger and emotions.

Forgiveness means accepting that no one owes you anything, not even an apology.

Forgiveness is extending mercy: hatred cannot be corrected by hatred; only love can heal.

Forgiveness is the path to true healing and abiding health.

Forgiveness is choosing harmony and peace over the desire of being right.

Forgiveness is giving yourself permission to feel joy and happiness again, knowing that you are not what happened to you, neither are you what people said, did, or did not do.

Forgiveness is setting down the baggage of the past and becoming lighter in your mind, heart, and body.

Forgiveness is the path to inner peace.

Forgiveness is reclaiming your power and knowing that what others did or did not do has no power over you or your life.

Forgiveness is freeing yourself from the bondage of the past.

Forgiveness is understanding that we all make mistakes sometimes and that no one is perfect.

Forgiveness is "acceptance". Giving the gift of acceptance to yourself and others and accepting the past just as it is.

Forgiveness is taking responsibility for the role that you have played in the situation and being open to learning life's lessons involved in the situation(s).

Forgiveness is a decision of taking back your power, knowing that no person, place, or situation has power over you: you are free and you are powerful.

Forgiveness is choosing the high road of love, no matter what!

Forgiveness is choosing the path of freedom and inner peace.

Forgiveness is deciding to free yourself from all emotional sludge so that you can allow abundance to flow in your life.

Forgiveness is being "mentally willing" to walk in the shoes of the other person and to be open to understanding where the other person is coming from.

Forgiveness is accepting to see the divine in others and seeing them as the innocent children of God that we all are in truth.

THE TRUTH ABOUT LIFE

Forgiveness is choosing the path of kindness and gentleness.

Forgiveness is deciding to reveal your inner angel through your willingness to choose peace and love.

Forgiveness is joining with the angels who rejoice with you every time you forgive.

Forgiveness is cleansing yourself from all past stories and hurts.

Forgiveness is cutting the cords that attach you to people, places, and situations and setting yourself free.

Forgiveness is deciding to turn your back to the past so that you can move forward in life and fully enjoy the present moment.

I have faced so many tribulations in this world—from being born to a father who cares little about me, being raised with the feeling of not being loved and accepted by my mother; being in abusive relationships; going through one of the nastiest divorces with my ex-husband going behind my back to empty our common bank account and leaving me without money and with nothing to survive; having my daughter taken away from me to a foreign country behind my back, leaving me in fear and wondering when I would see my daughter again; having been betrayed by the very people who I love deeply and trust and saw as my family but who later dumped me like an irrelevant piece of trash, etc. I lack words to describe all the suffering and pain that I have been through in this life. However, in the middle of the tribulations, in the middle of the pains, in the eyes of the storm of suffering, I came to find peace: I find peace in forgiveness. Forgiveness has been lifesaving for me. In fact, I have no other choice than to forgive as I have discovered that if I want to survive in this life, I must let go of all resentment and of the burdens of pain from the past. I aim to find peace, and I find it through forgiveness, meditation, and prayers.

When I tell my story to people—how I was healed from many diseases by doing a process of self-purification through forgiveness—the questions that I get asked over and over are: "What did you do to forgive them? I want to be healed too; how can I forgive everyone? What can I do to forgive those who have deeply hurt me and destroyed my life? How can I forgive the person who has murdered my loved one? What can I do to forgive that family member who had deeply betrayed me? How can I forgive my ex-husband?" To be honest, forgiving people was not easy for me either; it was more like a long journey. In fact, I have tried many things; I took forgiveness classes, did typing exercises, memorized Bible verses, used forgiveness mantras, said to myself that I have forgiven people many times, only to have the pain from past hurt and betrayals resurfacing in my mind and causing suffering. From my long years of forgiveness journey and exploration, I have found techniques, tools, and practices that have personally worked for me. By the request of others, I am putting together these techniques and methods that I have personally used and that have worked for me in an online course format that will be made available in the near future—to those who are interested.

Yes, Mahatma Gandhi was right: "The weak can never forgive. Forgiveness is the attribute of the strong." Indeed, it takes a deep strength to forgive. Yes, Mahatma Gandhi is an ascended master, and he came to teach us how to resolve our conflicts peacefully through the process of forgiveness.

The Truth about Life
CHAPTER 19

OTHER PLANETS, GALAXIES, AND BEINGS FROM OTHER DIMENSIONS

"Even if you are a minority of one, the Truth is the Truth."

–Mahatma Gandhi

Do Other Planets and Galaxies Exist, and Are There Beings on Them?

Yes, there are billions and billions of galaxies, and each galaxy contains billions and billions of planets. There are far more planets and galaxies than we know and that our scientists are currently aware of with the technologies that we have today. The best word to describe the immensity of the Universe and the amazing phenomenon that we all are part of is 'Infinity.' If you can see the whole Universe as the sea, the earth is like a drop of water of that sea compared to the immensity of the Universe.

Yes, there are beings that live on other planets. What would be the purpose of creating countless empty planets if all these billions and billions of planets were inhabited? The earth is simply a grain of sand among the immensity of planets and galaxies that exist in the Universe.

There are also other universes. Indeed, there are eight universes total, and there is even another universe that is currently being created. It is all about evolution and expansion.

Some planets exist on etheric forms, which means that they are not physical like the earth, and yet they exist, and there are souls that inhabit these planets and who continue their evolution journeys in these planets as nonphysical beings. Even in our solar system, there are planets that exist only on etheric forms. However, our scientists do not have the necessary technologies and equipment to prove the existence of these other planets yet.

The earth, our physical world, is created by God so that We (all of us who constitute God) can come and experience what it is like to live in physical form as separated beings and to learn to transform fear into love. How do you know what love truly is unless you get a chance to experience the opposite of love, which is fear, and compare them? It is a sort of adventure and exploration for us in a sense. We can tell by the diversity of planets and all the marvelous things in the Universe that we, as offspring or extensions of God, love creativity, growth, learning, exploration, adventure, and thriving for new things all the time.

What Can You Tell Us about the Pets and the Plants?

The pets and plants all came to serve humans and to ease their journey. I suggest to everyone to be kind to animals because their souls are pure, and they came to be in service to humanity. Even for the animals that we eat for food, blessing them with love, giving them natural nourishing foods, and raising them in the most loving environment is the best approach to adopt. Truly, these animals came to be in service to mankind, and because of our ignorance and greed, these animals are sometimes poorly treated, stacked in industrial factories, fed with hormones, and used for the purpose of making money. Thankfully, more people are awaking to love, and many farmers are now raising their animals with love and feeding them with natural healthy food.

An example of other animals that come to help and assist humans are the dolphins. Dolphins are beings from Sirius that came to help humans heal and teach them. Many people started to discover the healing power of being near dolphins or by listening to the sounds that they make. Currently, there are many dolphin meditations that people are currently using for soothing and healing purposes.

I would say even the pets came here to help humans and to bring joy and love in our lives and ease our journey. Pets are fulfilling their missions, as many people love their pets, feel comfort and joy in being in the presence of their pets, and deeply appreciate their presence. When you consider all of these things the Universe has given to ease our journeys here on Earth and help us to be at peace and happy, it is undeniable that we are deeply loved and really blessed.

What Are Your Thoughts Regarding the Amazing Progression of technology That We Are Seeing currently in the World Such as the Birth of the Internet, Computers, Facebook, and so on?

Our planet is shifting to a fifth dimension, and this change entails several aspects such as the awakening of people but also a technology revolution such as the one that we are currently seeing on Earth. The technological revolution in the last decades has been amazing. There is a divine time for everything, and the amazing evolution of the science and technology that is seen today in the world is a divine plan and orchestrated in a way that it supports the awakening, transformation, and the passage into the fifth-dimensional plane that is currently happening. Also, some of the genius and the great minds who are inventing these technologies came from a higher dimensional plane, and their purpose is to bring these technologies at this time and point of the evolution of the earth.

These technologies are helping to communicate with each other, no matter where they are in the world, but they are also supporting the awakening of the earth. As mentioned previously, some of the behaviors that used to be hidden such as injustices or other unloving behaviors can no longer be hidden and are being revealed or exposed, thanks to the new technologies such as internet and digital recording and social media. As the light is shining on the fears that have dimed this planet, darkness can no longer hide, and humanity will have to address them and find a way to heal its issues. Also, the work of the lightworkers, star seeds, and other beings coming from higher dimensional planes that are here to help people to awaken will not be possible or will be very challenged without the presence of new technologies such as the internet and so on.

Some may argue that there are inconveniences associated with these technologies, and perhaps they are right to some extent, but overall, everything is pointing in the direction of the evolution and happiness of humanity. There is a bright future for this planet, and this is only the beginning. I am not saying all the genius or the great inventors are from higher dimensional planes or they have something more than others because we are all geniuses and inventors by our divine nature, and we all have the same potentialities in spiritual truth, but it is true that some inventors of these new technologies are coming from other more advanced planes and dimensions with the purpose of bringing new technologies to help humanity, and this is good news.

Is It True That There Are Angelic beings who came on Earth and

Helped in Various Fields such as music and so on?

Yes. Some of the great composers and great musicians of the classic history of the earth came from the angelic realm to help. The truth is that the earth is a 'salad bowl,' and we all come from other places. Yes, there always have been angelic beings among us, they have always come, and they will continue to come to help in the future. These angelic beings are like everyone else. They came to serve and bring a light and love not just in the music field but wherever they are, whether they are flight attendants, nurses, waiters, or singers, and so on.

The Truth about Life

CHAPTER 20

HEALTHY LIVING, FOOD, EXERCISING, SUN, GROUNDING, AGING

"Truth is ever to be found in simplicity, and not in the multiplicity and confusion of things."

—Isaac Newton

Food and Physical Activity

What Is the Truth about Food and healthy eating?

The truth about food is simple: you become what you eat. You will start to feel like what you eat. If you wish to feel like a processed hotdog, then you can eat a processed hotdog, and you will embody the processed chemicals and the mixture of salt and fats, and you will start to feel that way. Therefore, decide what you want to feel like and eat accordingly. You will notice that plants and foods that are in the sunshine exude happiness. Eat happy food. Eat fresh and real food, and that is what you will become like. It is simple. These are basic concepts for living in high vibration.

So much have been written and said about foods and healthy eating. There is a whole industry developed on this. There are weight loss programs, healthy eating coaching programs, calorie-counting diets, Paleo *Diet*, Vegan *Diet*, Low-Carb *Diets*, Dukan *Diet*, Ultra-Low-*Fat Diet*, Atkins *Diet*, HCG *Diet*, Zone *Diet*, etc. You name it; you will find a diet program about it. However, the truth is simple, much simpler: you become what you eat. Every food has a specific vibration and frequency that can be measured in hertz. In fact, there is scientific equipment nowadays to measure the vibrations of foods. Fresh row fruits and fresh vegetable, in general, have the highest vibration.

You can see the process of eating as transferring energies from these produce to your cells, and the nutrients and life force of these produce will invigorate and energize your cells, which will reflect the health and aliveness of the fruits and vegetables that you have consumed. Therefore, when food is overcooked, dried, or unnatural, there is no aliveness in it since it is dead food. Moreover, not only do processed foods not have any aliveness or life force in them, but they also contain chemicals, fats, salts, and all kinds of things that you are ingurgitating in your cells and filling your cells with chemicals and toxins, which are detrimental to health. Therefore, if you want to feel alive, increase your vibration, and want to have vibrant health, then eat healthy food, happy food, natural foods, foods that are filled with aliveness, and foods that are light. One of the challenges that many people face is that they want to eat healthy, but they have all kinds of cravings. They crave sweet, sugar, and processed foods, even though they desire to be healthy, but there is a way out of the craving.

What Is the Truth about Craving?

Craving can be due to several things. First, it may be that your body needs some nutrients or vitamins that it hopes it will find in some foods, so the body will crave these foods. Keep in mind that foods that are cooked or boiled, baked or processed often have no good nutrients left in them, as they are destroyed by the high temperature of the cooking process. Many live their entire lives and barely eat any natural fresh fruits and vegetables; therefore, it is really possible that the body may not be receiving the nutrients that it needs.

Another important thing to know about cravings is that there are etheric cords, energetic cords, that are unseen by the physical eyes that link people to the things that they are craving, and this is one of the reasons why they go back to what they are craving again and again, and they can't simply stop desiring them. It is the same principle as addiction, and that is why addiction is so difficult to stop. Interestingly enough, the term 'food addiction' is now being used to describe the phenomenon of craving.

How this works is that anything that you engage in goes stronger.

Let's take the example of a hypothetic person that I will call John who comes across chocolate for the first time in his life. Before that first time, John has never craved chocolate, and how can he crave chocolate anyway since he never tasted it and doesn't even know what chocolate is like? One day, for the first time in his life, he came across chocolate and ate some. It was like an explosive blend of taste, sweetness, in his mouth, and he loved it. Chocolate became one of his favorite snacks, and he started buying and eating it in on a regular basis. An affective link now ties him to chocolate; all the sensations and the good feelings that he had every time that he ate chocolate are all contributing factors that further reinforced and attached him etheric-wise to chocolate. On an energetic level, there are etheric cords that developed and linked him to chocolate, and these cords are the basis that fuels his craving and what will lead him to go back again and again to chocolate. This is true for other addictions as well.

Also, people who are in abusive relationships and who have a difficult time leaving the relationships, most of the time, there are etheric cords that have developed with time and that link them to their abusive partners. That is often the reason why they have trouble leaving their partners, or sometimes when they leave, they're still thinking about their partners because the cords have not been cut, and they are still attached to their partners. Many people falsely think that those who have cravings either can't simply control themselves or they love food too much or that they are weak and so on. However, this is not the case because people who have addictive cravings to foods simply have etheric cords attaching them to these foods, and that is the reason why they have trouble releasing these foods and why they crave these foods. Cutting these cords will help them to detach from the foods so that they can free themselves from these cravings. Again, the same is true for relationships as well. When we truly understand the principle and the truth about cravings, we understand that there is no need to judge others who have cravings or ourselves. There is no need to feel guilty about anything or to blame yourself because you have cravings or have trouble not eating some food. This is why we must stop blaming ourselves and others and instead have compassion on ourselves and others because there are other factors that are involved that we may not necessarily be aware of. Therefore, we must stop judging or criticizing others because they are overweight or obese.

Light of Truth

I used to have cravings for various foods and was basically addicted to many foods, candies, chocolates, sweet foods, caffeine, and so on. I have tried so many things unsuccessfully. I have tried various diets, calorie counting, and even took weight loss pills at some point in my life to help keep my weight and let go of the craving. I spent many years of my life not knowing how to deal with my weight issues and my addictive craving to sweet foods.

Nothing has worked, or at least they worked for short periods of times only; then I would fall back to the old ways of addictive cravings for foods and weight issues. I was doing a yoyo with my weight and felt like I was trapped and kept spinning my wheels in a vicious endless circle of weight loss and weight gain. After learning the truth about craving, I decided to ask for help by asking the Archangels Michael and Raphael to help me heal from my craving and to cut the ties that liked me to these addictive substances and foods.

Those who are familiar with Archangel Michael know that he carries a sword; in fact, one of the purposes of his sword is that he uses it to cut cords that bind us to foods, people, and situations. With that knowingness, I started to ask for his help and the help of Archangel Raphael, the healing angel, to cut the cords that linked me to the foods that I craved. I did this mostly through prayers. I was skeptical at the beginning because I did not know how the many years of addictions to food and to caffeine could be gone, and I didn't even know that there was hope for me, but I decided to try to pray for the cords to my craving to be cut.

I wish I could say that I woke up one day, and all my cravings were gone, but that was not my case. It was a steady and yet an undeniable and effective process for me. I let go of the caffeine to my great amazement. One day, I started to notice that I did not crave chocolate anymore and could go to the grocery store without having the urge to buy candies, chocolate, or cookies. I was amazed. "It is working," I said to myself to my grand astonishment. This gave me hope and the motivation to keep praying for help to heal from craving because at that point, I decided to become a vegan, and I still had craving for French fries, roasted Mexican chicken (namely, 'Pollo a la Brasa'), as well other cravings for pizzas, tasty seafood, and so on. Slowly, I started to feel a real change, and the cravings were scrambling away one by one. In about a two-year period, the craving almost completely disappeared. It was almost like a miracle for someone like me who suffered from decades of addictive cravings to foods and struggled with weight issues to be healed from cravings. I can tell from my own experience that asking for help to cut the cravings startlingly works and that these etheric cords are real.

The only thing that I want to add is that you have to be willing to release the cravings before the angels can fully help you. I am saying this because I was reluctant to release some of my cravings, especially chocolates, ice cream, coffee, cookies, and seafood, because I felt like I would be missing out on something and was not willing to let them go, even though a part of me wanted to heal from craving. Therefore, I had to come to a point when I said to myself, "Fine, I surrender them," and sincerely decided to let go of them and ask for assistance.

What Can You Tell Us about Exercising and Physical Activity?

Nothing is static in the Universe, as everything is dynamic and in constant change and movement; this is so for our physical bodies as well. The body needs to be in movement; in other words, we need to exercise the body every day. In fact, when we exercise, the body releases toxins and negativities and also produces 'good feeling' chemicals and hormones such as endorphins. That is why people feel great and happy after exercising.

The main thing regarding physical activity is to find an exercise that is pleasurable and enjoyable for you. It has to be something that you love and enjoy; otherwise, it will not be sustainable with time, and you will not continue over a long period of time. What often happens is that many of us choose physical activities that we do not love or enjoy or think that we have to go to the gym, and it becomes a chore and a pain for us, instead of being a joyous activity, thus making it difficult to sustain with time, and then we give up. However, if we can find a physical activity that we love, it becomes more of a pleasure and a joyful activity rather than a chore. When we exercise and sweat, the body releases toxins, thus detoxifying not only from chemicals but also from mental and emotional negativity.

There are also many other scientific and biological benefits and mechanisms involved in exercising but also energetic and electromagnetic changes that happen in the body and in our energetic fields when we exercise. Also, when we do physical exercises such as running, it clears the chakra systems, leading to several health benefits. There are, of course, other benefits related to exercising on the physical level but also on the emotional, mental, and energetic levels.

The main and perhaps the biggest truth about exercising is that we don't need a lot of time as taught in society; 15 or 20 minutes a day of good intensive physical exercising is enough to maintain a healthy balanced body. I was shocked the first time that I discovered that we don't need that much time, and I have tried the 15 minutes of good exercise a day, and it works amazingly well for me. Of course, some physical activities such as walking may require more time, but overall, we don't need a lot of time. I will say to go with what works best for you, and if you feel like your body needs 30 minutes of exercising daily, then go with that. I will add that it is preferential to do the exercise until you sweat. However, it has to be a daily activity and practiced on a daily basis. Just like brushing our teeth is part of our routine, exercising should be perceived in the same way: an essential routine that should be incorporated in our lives.

Tips and Techniques for Healthy Eating Habits to Increase Your Vibration

We all know the importance of eating healthy and the health benefits associated with it, and yet there is more to be learned and discovered on that level. The truth is, every food has a specific vibration. Fresh produce such as vegetables and fruits have the highest vibration, especially when they are raw. Produce that is grown under direct sunlight also has one of the highest vibrations. Eating these high vibration foods will help you to keep your vibration high as well. The truth is that every person is unique and will have a diet that is specific to their needs. Vegan foods have the highest vibration. When I say vegan foods, I mean the fresh natural foods that are not processed. However, everyone will have to find what food is best for their needs and their life purpose. Our world is changing, and its vibration is increasing, and more people will opt for vegan and organic foods in the future as we move along.

Tips or Practices for Healthy Eating

<u>Eating light and in light</u>

Eating light and fresh produce: Fresh fruits and vegetable have the highest vibration and will help you to keep your vibration high. When vegetables or fruits are cooked, dried, processed, or baked, they lose much of their vibration. Fresh <u>raw</u> produce is the best and has the highest vibration. When you eat light food, you will become lighter in your body and also in your mind. Eat light because we are light beings.

The higher your vibration, the healthier foods you will crave. As you increase your vibration, there may come a time when you will crave only salads, vegetables, and fresh fruits because your vibration will become so high that your body will crave only healthy food. This is a truth that many do not know, but indeed, "health attracts health." Therefore, the healthier you get, the more healthy foods you will desire, which will further boost your health.

<u>Eating organic foods</u>

Organic foods are better since they have fewer chemicals and toxins. For those who cannot afford organic, you can sincerely pray over the food before eating it, bless the food with love. In fact, it is best for everyone to seriously and sincerely bless their food before eating it, whether the food is organic or not. All I will mention here that prayer does make a real difference, and it increases the healing ability and the vibration of the food. Also, there are guardian angels and spirit guides around us at all times who help purify food and increase its vibration when you pray over your food. Below is an example of a prayer to bless your food:

"Dear God, thank You for blessing these foods. Please fill them with healing, light and love." (*Visualizing the food surrounded by bright white light is very powerful, but if you cannot visualize, it is okay too; your intention is enough.*)

Purifying your water

Finding a great water purification system to purify the water to remove fluoride and other toxins and chemicals is essential. The water coming from our tap and water system is filled with fluoride and other chemicals that are detrimental to the body. Purifying the water helps remove the chemicals and will help you to keep your vibration high. If you cannot afford a water purification product, you can simply pray over your water before drinking it. Water is curative when prayed over before drinking. Indeed, I do not drink tap water and purify all the water that I use in my house. I also stopped drinking prepackaged water in plastic bottles found in grocery stores years ago, as these often contain BPA (Bisphenol A) and other chemicals. BPA is an industrial chemical that has been used to make many plastics. I purify all my water and also pray over my water before drinking it.

Another important truth about diets is the importance of drinking enough amount of water. In today's society, we barely drink enough water, and sometimes we even replace water with sodas and other drink, and that is unfortunate because the body is mainly made of water, and water is essential for the body's cellular mechanisms and to maintain a healthy body. I personally try my best to drink approximately six liters (1.5 gallons) of water daily. I discovered that drinking a lot of water helps improve my health, and it also helps eliminate chemicals in the body via urination. Furthermore, I stopped drinking sodas and only drink water. Drinking enough water is a component of the recipe for increasing one's vibration and keeping it high.

Juicing

Juicing is also a great approach to adopt, as it helps to consume a great variety and amount of fruits and vegetables at once. It truly helps to detoxify the body from chemicals while nourishing it with healthy elements and vitamins.

Being present with the foods while eating

Not thinking about other things and not being lost in thought while you are eating is also a healthy habit to adopt.

Sunlight, Electronic Wavelengths, Grounding

What Is the Truth about Sunlight?

Just like plants need the sun and water, so does the human body. The body needs to get sun rays daily in order to perform some vital functions. In our modern societies, we spend a lot of time indoors, and we barely get the amount of sunlight that our bodies need. Nowadays, researches recognize many benefits of the sunlight such as its essential role in the synthesis of vitamin D that is so indispensable for many cellular mechanisms and functions in the body. For instance, many people have discovered that they are happier and less depressed when they spend time outdoors under the sun, but they don't know why. Spending time in the sun does indeed help improve our mood. This is the basis of light therapy or luminotherapy.

Sunlight is even more essential for our wellbeing than we know. In fact, the body absorbs sun rays and transforms the energy of the sun to support vital cellular and energetic functions of the body. For instance, if we compare the body to a computer that needs to be charged, the sun can be seen as an 'energetic solar system supplier,' and being in the sun daily is similar to recharging the battery of your body with the vital energetic force that it needs for functioning. If you do not plug and charge your computer frequently, its battery will go dead, and the computer will stop functioning. Similarly, if you do not plug and recharge your body with the solar energy source that it needs for its function, the body will be depleted in energetic force and will not function at its maximal capacity.

The truth is that the body needs sunlight, and the sunlight is vital for the body in the same way that it is vital for plants. We have been told that the sun is detrimental to the skin and that it is the cause of many skin-related diseases, which is not true. We have been told that we need protective and other sun blockade creams in order to protect the body and the skin from some 'defecting rays of the sunlight,' which is largely false. We have been told that we need to protect our eyes with tinted sunglasses in order to have healthy vision, which is untrue. All of this misleading and untrue information has probably originated from specialized industries and people that benefit from this misinformation by selling their products. This untrue information has somehow become ingrained in our minds, become accepted by the general population, taught in schools, supported by the healthcare community, and believed and accepted by the whole world as truth. The truth is that it is not the sun that causes skin cancer or whatever diseases are attributed to the sun. If it was the case, the majority of Africans who live in the sun with their bare skin with no protection and who don't even know what sun protective creams and products are should all have skin cancer, but that is not the case. Furthermore, the chemicals in the sunscreen protection products are even toxic and detrimental to the skin. I will not expand further on this topic but only reiterate the importance of daily exposure to the sunlight and its vital functions for the healthy functioning of the body.

Light of Truth

Somehow, the first time I was presented in class with the so-called detrimental effects of the sun in our bodies and that the sun is the cause of disease and that we need to protect ourselves from it, I intuitively knew that it was false information. Deep inside, I knew that it was not true, but I was in school and needed to study the material in order to pass the tests and earn my degree. However, I never bought into the lies, never use sun-protective creams when I am in the sun, and do not use sunglasses. Years later, I have learned from other scientists about the benefit of the sunlight and the importance of being in the sun daily. Therefore, I have started to go in the sun daily for 30 minutes or so and started to meditate and pray while I am in the sun. I discovered that being in the sun and doing my ritual meditation and prayer was improving my health, and I was energized and filled with light literally and figuratively.

What Are Your Thoughts about Exposure to Electronic Wavelengths and So On?

It is true that exposure to electronic devices and electromagnetic waves is not good for health in general because of the bombardment of the wavelength frequency on the body. However, the most damaging things for the body are low frequencies coming from negative thoughts, anger, resentment, and other negativities of the mind. Loving thoughts and a loving environment are healthier for the body. However, reducing exposure to electronics by staying away from electronics and spending time outdoors is a good approach to reduce this exposure.

Grounding as a healing solution for reducing electronic exposure

Grounding is a natural way that helps to decrease the wavelength of electronic bombardment that we are subject to in our everyday lives. The earth has a natural electromagnetic vibration that helps in healing. Grounding is a natural technique that consists of plugging into the natural radiation of the earth by walking on the earth barefoot or swimming in the sea or hugging trees that are grounded on the earth. The soles of the feet have to be in direct contact with the natural ground of the earth.

What Can You Tell Us about Music and the Effect of Music on the Body?

Soothing music has a healing effect on the mind, body, and feelings. The molecular properties of music can soothe and shield. The most important thing is to pay attention to how a particular piece of music makes you feel. Does the music make you feel good and happy, or does the music seem irritating to your ears? Do you feel like dancing and smiling when you hear the music? These are some questions to ponder that can help you to know if the music is healthy for you or not. I would add here that using feelings to know whether something is good for you and is in your best interest applies to anything and everybody that you will encounter in life and not just for music. Again, it is all a matter of vibration. People are becoming aware of the properties of music. Nowadays, sound healing and the science of quantum physics have started to explore these new avenues of healing.

The Truth about Life
CHAPTER 21

MONEY, MANIFESTATION, ABUNDANCE

"No legacy is so rich as honesty."

–William Shakespeare

The Truth About Money

What Can You Tell Us About Money?

The topic of money is an important one, as it concerns all of us. Money itself is just a means of exchange of services. However, money represents freedom, and what we are all seeking in fact is the freedom to do whatever we want, the freedom of being able to buy whatever we want, freedom of not being constrained to go to a workplace that we don't like, freedom to travel and visit the world, freedom of not having to clock out and clock in daily, the freedom of not having to wake up at a fixed time every day, the freedom of being able to spoil our loved ones with gifts and make them happy, the freedom of paying all of our bills and all our needs on time with no fear ... freedom. In fact, we are free beings, and freedom is rooted in the very core of our souls, and that is why anything that is restrictive or that limits us does not feel good, and that is why we are constantly seeking freedom.

What we are all seeking is freedom, love, and peace. However, we have been taught and conditioned by society to believe that money is what we need or what will bring us peace, love, and freedom that we are seeking. In fact, you can have money and be very miserable and sad. We have seen many stars and famous people who have money but live a very pitiful life, feel lonely, are very unhappy, and sometimes even commit suicide. Indeed, you can have money and not have any freedom at all, and sometimes many sacrifice their freedom in order to get money. For instance, I know physicians who made around half a million dollars a year and yet work nearly 16 hours a day and never truly have the time to enjoy life and are really stressed out, not free, and sometimes very unhappy. Many people work five days a week, for long hours, in order to get some money and never have the freedom to live and enjoy their lives. We give up our freedom in order to get money, while in truth what we are truly seeking is freedom. What a paradox!

The truth is that sometimes these jobs are very restrictive and do not give us any autonomy and the freedom that we are seeking. Many jobs are restrictive in nature, where you have to clock in at a specific time every day, having only 30 minutes of break to eat, and sometimes many do not even have time to take a break or even eat. On top of these restrictions, the house loans, car loans, school loans, and the credit cards that we have in the United States set most people up as 'prisoners of their jobs' for life. How shocking is it that many of us take on long-term loans of 30 years to by our houses? It is similar to having your hands tied for life. There is no freedom in that, and yet society convinces us or we convince ourselves that these loans and these restrictions are the path of freedom.

It is heartbreaking when I see people doing the same thing all their lives and going to workplaces where they are unhappy because they believe that there is no way out or because they believe that it is the only way that exists. The truth is that we can manifest any life that we desire, and there is a way out of the restrictive prison-like corporation systems. There are ways to make money and to be abundantly cared for in this life without losing our freedom. However, this requires having faith and being able to take risks, and sometimes, that will even require starting your own business. Many do not even believe that it is possible; they think it is too risky because they want the security that comes from having a paycheck; they are afraid of failing or think that success is reserved for only some fortunate people. However, is that even 'security' if you are not free to do whatever you want or when you are sometimes even constrained to do things that you do not believe in or that you dislike just to have a paycheck?

The truth is that we are free beings, and we have a defined life purpose for our earthly life. When you are in your life purpose or doing what you came on Earth to do, things naturally flow easily and are more enjoyable for you, and money will come to you as well in the process. However, the key is to know what this life purpose is and what are the gifts and talents that you are equipped with. Many do not even know what their life purposes are and simply take the first job that they find, and they are so anxious for the financial situations that they are afraid of taking any risk. However, every successful person has taken risks to be where they are today, and the only guarantee of not succeeding is not having the courage to try. It takes a great deal of faith, courage, and strong determination to get out of the restrictive system and find your own path.

THE TRUTH ABOUT LIFE

Money is a very good thing, in fact, and it is God's will for everyone to be abundantly cared for. Money itself is just a means of exchange of services. For instance, if your hairdresser provides you a service with her skills and talents, in exchange, you will give her money for her services. Like everything in the Universe is energy, money itself is energy. Money is not good or bad; it's simply energy. It will take on whatever significance or meaning you assign to it and what you believe it to be. What often happens is that many of us want to have money, but at the same time, we unconsciously have negative thoughts or limiting beliefs about money that drive away this same money that we desire. For instance, someone would love to have money, but at the same time may have the belief that money is the root of all evil or may be jealous of those who have more money than him or think that there is not enough money for everyone in this world or may believe that he has to work hard to earn money or may simply believe that he is poor and will always struggle with money. All these negative thoughts and feelings and beliefs about money will only keep him in a state of poverty because the Universe is always listening and responding to our thoughts, feelings, and beliefs. This is because the Universe works according to the laws of vibration or the law of attraction, and money is not an exception to this law.

The truth is that the Universe is abundant, and there is enough wealth and abundance for all of us; however, everyone has to manifest the money according to the law of vibration or attraction, and no one is exempt from this law. In fact, the Universe is very fair and does not do favoritism. What often happens is that we create and manifest our abundance in life via our thoughts, beliefs about money, our capacity to believe in our own ability to succeed, and our positive outlook and faith that it is possible to be abundant. The more faith and trust you have, the more you will get. The more you believe that you are deserving of money, the more money will come to you. There is a full recipe for how to manifest money and abundance, and more people are rediscovering that currently on Earth. Deservability, worthiness, faith, positive thoughts, happy feelings, and joyful emotions are elements that play important roles in how much money will come to you. The truth is that the Universe is unlimited and constantly expanding; therefore, someone having a lot of money is not taking from anyone else, as there is enough abundance for all. Many people have some deeply rooted seeds of unworthiness, lack, and other outdated beliefs that led them to stay in positions where they do not have enough money because of limited belief systems. However, these beliefs can be changed, and as you change your thoughts and beliefs, you will open the way for more money to come to you.

Steps for Manifesting Abundance and Money

How Can We Manifest Money and Abundance?

First, I would like to clarify that abundance is not just having money. Abundance in its truest form comes as love in relationships, health, joy in career, and wealth in money. There is a full recipe for how to manifest abundance. Since we create everything with our thoughts and feelings, manifesting abundance also comes with changing our thoughts and feelings. It is an inner work, and it starts from the inside and then manifests on the outside as abundance. In other words, you will have to literally become abundant inwardly first. To be abundant, you have to think thoughts of abundance, prosperity, and being in a vibration that allows abundance. However, by following some steps, you can steer the manifestation process in the direction that you want. It does not matter where you start from because anyone can change, apply these principles, and manifest the abundance and the money that they desire in life.

THE TRUTH ABOUT LIFE

Below are some important key elements for manifesting abundance:

Gratitude:

Feeling gratitude for what you already have and maintaining a gratitude journal is really powerful and helps considerably in manifesting abundance. This is because when you are grateful for what you already have, more will be added to you. This is a law of the Universe: if you are grateful and thankful for something, the thing will expand. For instance, Oprah Winfrey, who is among some of the healthiest people in our country, stated that she maintains a daily gratitude journal and that her abundance is related to the fact that she is always grateful for what she has.

Being grateful does change your vibration, uplift your mood, and bring more abundance into your life. The question that remains is, how can you do this? I will suggest that you find a nice notebook that you like, or you may also use your notepad on your cell phone. Every day, write down at least three things that you are grateful for in your notebook. For instance, you can write, "I am grateful for the nice bowl of rice that I ate this afternoon," "I am grateful that my children are healthy and happy," "I am grateful for the great night of sleep tonight," "I am grateful for talking to my best friend today," etc. You can write down anything that happens during the day that you are grateful for. The more you write these gratitude affirmations, the more things you will find to be grateful for. At the beginning, the exercise may look odd to you, especially if you have never done this before, but there will come a point when you will have more than a dozen things in the day that you are grateful for. Doing so will change your vibration in a positive and high vibration that allows abundance.

Visualization:

This step consists of visualizing your goals as if they were already done with positive feelings. Let's say that you want to manifest money and your dream house. You can visualize how it feels to have all your debts paid and how it feels to buy your dream house and the joy and the excitement that you experience while you are in your house enjoying sweet tea in your backyard and so on. You may visualize what the house looks like in vivid detail and the joy of having that house. You can create in your mind whatever you desire and visualize it as if it is already a real and done deal. There are awesome manifestation visualization meditations that can help you tremendously with the visualization step. I will suggest that you find good visualization meditations and practice them daily or as often as you can, or you can also refer to the suggested meditations at the back of this book to help you in this process.

Another important tool that tremendously assists in visualization is making a vision board. A vision board is a board that you create with pictures of the things that you desire to manifest and hang it somewhere in your house or bedroom where you can see it every day. Do not be fooled by the simplicity of this exercise because when you create the vision board, your energy and focus will be on these images every time you see the vision board, and with time, it is as though you have one foot already there. It works. I have experienced this firsthand, and I can attest that vision boards do work, and you will be surprised to see the material things come into your life one day. In fact, Jack Canfield, the author of *Chicken Soup of the Soul*, recommends people make a vision board and teaches how to make them. These successful people know the secret of how to manifest abundance and strategically apply these principles in their lives to succeed. However, it is something that is within the reach of everybody, and everyone can do this.

The key to creating a vision board is to choose pictures that are really attractive to you and that represent what you really want to create. For instance, if you want to buy a new home, find a nice picture of your dream home and put it on your vision board. Do not choose a picture of an old home or one that you do not like 100% because you don't want to inadvertently create something that you don't really want. You can also lookup how to create a vision board on the internet to assist you with your visualization process.

Having clear defined goals:

In this step, you are asked to clarify for yourself what you really want to manifest and set clear intentions. How can you manifest anything if you don't even know what you want or have a clear and precise idea of what you want to manifest? Let's suppose, for instance, that you want to build your own house, and the house developer asks you what you want to build, and you tell him, "I just want to build a nice house." He may not help you appropriately and will need further details and precisions in order to build the house. Your house developer will probably ask you the following questions:

What kind of house do you want to build? Do you want the house to be built with wood or stone? Do you want three bedrooms or five bedrooms? Do you want the house to be all built on the same floor, or do you want to have stairs? Which size do you want the house to have? And so on. Similarly, you have to know or at least have a clear idea of what you want to manifest and set your intention for it before you can manifest it. You do not have to know 'how' you will manifest it, but you have to know what you want to create. I will suggest that you find a book that you will call your 'manifestation book' and write down the projects, the goals, and your intentions of what you want to achieve with precision.

In fact, I have a book in which I write down my goals and the things that I want to manifest, and it always amazes me every time I come back to my book a year or months later and realize that many of the goals that I wrote became true and have manifested in my life.

After writing down your goals, you may want to write some intention statements or affirmations for each goal. Here are some examples of intention statements that you may use and tailor to your goals:

It is my intention that I experience only vibrant health that will support my life's purpose.

It is my intention that I will start my business.

It is my intention that I will become a physician, a CEO, a mother, or a business owner, etc.

It is my intention that I live a peaceful, fulfilling, and inspiring life.

It is my intention that_____ (fill the blank).

Affirmations:

This step consists of writing down affirmations of your goals and saying these affirmations daily. You can also record these affirmations and listen to them daily. You can also write down these affirmations daily. I personally found affirmations to be really powerful, and they will manifest sooner or later when you practice them. Affirmations are so important because they immediately connect you to your dreams and shift your vibration. Sometimes, especially when you are going through challenges and if your mind is filled with worries or simply filled with negativities, listening to your affirmations may perhaps be the only time during the day when you are thinking about something positive, about your dreams, and dreaming about a positive outcome, a better future, and doing so will automatically put you in sync with your dreams.

The truth about manifestation is that you have to connect to your goals on a regular basis in optimistic ways and dream that it is possible to attain them. That is what affirmations do; they enable you to experience your goals as though they are already done, even if it is not the case yet. Since the mind does not know the difference between what is real and what is just imagination, affirmations in a sense trick the mind and make you feel as though you are already living your dreams, and that is why they end up manifesting in the physical world as real objects, money, businesses, and experiences.

Faith:

Truly believe that it is possible to reach your goals and that you are able to attain them. The more you do the affirmations and repeat them daily, the more your faith will grow. This is why people call affirmation autosuggestions. Therefore, it is 'a positive reinforcing circle.' Practicing your affirmations will boost your faith, and the more your faith grows, the more motivated you will be to practice these affirmations and the more positive results you will get, which in turn will further boost your faith since you are now seeing the positive results of your efforts.

Surrendering control:

Let go of the desire to control everything, especially 'how' your dreams will come to life, and trust that a door will open. You will not know which door will open; you will not know when it will be open and how, but your only job is to let go and trust in the process. The only thing that will be required of you is to let go, surrender control, and trust. Leave everything in the hand of Universe, in *God's hands*, as many like to put it. However, letting go does not mean being passive and not taking actions; instead, it means being still and listening to your intuitive guidance and taking action when action is needed. This leads us to the next step, which is listening to your intuition.

Listen to your intuition and act upon it appropriately:

One of the most powerful guidance systems that we have been given is our intuition. The Cambridge English Dictionary defines intuition as "an ability to understand or know something immediately based on feelings rather than facts." Intuition, feelings, gut knowingness, hunch, sixth sense, following your heart, instinct, etc., all mean the same thing. Unfortunately, the words 'feeling' and 'intuition' have been banished from workplaces and from society because they are deemed untrustworthy, inaccurate, and people prefer to trust the facts and their logical minds.

Here is how it works: at any given moment, you have the voice of the ego that always comments, analyzes, compares, rationalizes, pretends to know, and will come up with ideas and many thoughts about a situation. At the same time, you have your intuition or feeling or that 'small voice' that will give you insights and clues about the situation. One of these two guidance systems is true, and the other is false and confusing. The guidance that is true is your intuition or your gut feeling. The guidance that is confusing and often false is the voice of the ego, the thinking mind, the analyzer, the commentator, and the rationalizing machine that pretends to detain the answers and creates doubts and confusion.

The intuition and your feeling is what is real, and it is the voice of your

THE TRUTH ABOUT LIFE

Soul (your true self), the voice of your Spirit Guides and of your guardian angels who are guiding you during your earthly experience. Learning to listen to your intuition, trust your feelings, and act upon them is the key to success in life, success in business and making money.

What many people do not know is that the most successful people in the world succeed by following their intuitions. For instance, Richard Branson, the founder of Virgin, a multibillionaire, who is one of the most influential people on our planet, said, "I can tell you that when I have to decide whether or not to go ahead with a new venture, I have often found that intuition is my best guide." He admitted relying on his intuition for business decisions.

Even Bill Gates who is one of the most successful people of our time once said, "Often you have to rely on intuition." These successful people have learned to trust their intuition and know how to use it to their advantage.

Oftentimes intuition will come in various forms such as a feeling, a knowingness, a sensation that something is off or out of line, a gut feeling, an 'aha moment,' a drive to take an action, a vision, a dream, a small nudging to go in one direction or not to take a path, etc. All of these are the voice of truth that is guiding you. What often happens is that many times, the ego will come later and fill you with doubts and let you second-guess your feelings. The ego will tell you, "This does not make any sense. No one does that. Are you sure you want to do that? This is not logical. You may fail by taking that action," and so on. However, to be truly successful in business and in life in general, you have to rely on your intuition and trust it and know when to take action. Oprah Winfrey, who is another influential woman of our time, publicly admitted that she relied on her intuition for many decisions in life and advised people to do the same. Your intuition is your best friend, your guide, and it is the voice of truth. Therefore, the key to manifesting abundance is learning to listen to your intuition, heed your feelings, and act upon them when necessary.

I remember that when I was in college, I often relied on my intuition to take tests and exams. Many times, when I didn't know the answer to a question, I often closed my eyes, scanned the choices given, and picked one answer based on my intuition and feelings. I often got it right and got good grades and was in honor rolls throughout my college curriculum. Of course, I studied a lot too, but honestly, my success was based not on my effort alone but also by relying on my intuition to help me to make decisions. I remembered that sometimes my friends with whom I studied in college would ask me how I knew that a specific choice was the correct answer for a question of an exam or a test, and all I could tell them was, "I just feel it." I realize today that I am not the only one using my intuition to guide me and that countless successful people also rely on their intuition for business decision-making, and that is how they succeed in life.

Forgiveness:

Forgiveness clears the path for abundance to flow to you. Unforgiveness and any other negative feelings only block the flow of abundance. Remember that manifestation is the 'art of positivity,' and thus, when you forgive, your mind is clear, and you can focus on your goals. When you do not forgive, your mind will be filled with resentment and hurt and thoughts about those who have irritated or hurt you, and there will be no space left in your mind to think and dream about your projects.

The arguments in the mind and unforgiveness are energy-draining and

THE TRUTH ABOUT LIFE

will deplete your energy, while a better use of your time and mind will be to devote them to your dreams and projects. However, many people have their minds filled with negativities and spend their day ruminating frustration and anger about little things such as a waiter who was unkind to them or a coworker who was mean to them, or they're filled with thoughts about past hurt or worries about the future. If the mind is filled with these negativities, there is no place for the positive thoughts and joyful feelings, which are the platforms from which abundance grows. Since anything you focus on will expand, you must first focus on thoughts of abundance and your goals in a positive and happy way for abundance to flow to you. This is the trick of manifesting abundance, and you must become the captain of your mind and focus on your goals and find something to be grateful for because when you think that you are poor, or when your mind is filled with unforgiveness, anger, and worries, it will push away abundance, and it often becomes a vicious circle. This is why forgiveness is so important. It affects not only your health but also your abundance and happiness. Therefore, the solution is to find a way to forgive everyone, to focus on abundance and positivity, no matter your situation.

Notice that those who are rich spend most of their time thinking about their wealth and ways to increase their abundance or where they will find the next big ideas for their businesses and so on. On the other hand, those who are poor spend most of their time thinking about lack and scarcities, how they are going to pay their bills, how poor they are, how awful their jobs are, those who have hurt them, how victimized by life they are, or being envious of those who have more money than them. Thus, the vicious circle continues on both sides, leading those who are rich to become richer and those who are poor to stay poor or become even poorer without knowing that it is their thoughts and feelings that are contributing to their situations. "For whoever has will be given more, and they will have in abundance. Whoever does not have, even what they have will be taken from them" (Matthew 25:29).

Taking action steps:

Taking small actions steps on your goals daily, even if it is 15 minutes daily, is very important. Just as when someone desires to build a house, even if the person does all the planning, visualizes the home, draws the house on paper, and affirms a thousand times a day "I am thankful that I have built my dream house," at some point that person will have to gather the physical materials such as the wood, brick, the pillars, the roof and put them together in order to have the house. In other words, you can visualize your goals, do the affirmations, dream about it, write them down, be positive, etc., but at some point, in order to see your dreams become a reality, you will have to get handy and work on them. This is just the truth of how it works. I have talked to many people who have big dreams in life, and they all tell me that they want to reach their goals; however, they have never taken any single concrete action step to transform their goals into reality. Many times, they just keep talking about their wishes and dreams for years and stay in this position of wishing something better for years and years. This is why 'taking action steps' is important and has to be done at some point to keep the goals and the dream moving forward, even if it means 15 minutes a day.

Here is how manifestation goes: "Think it, dream it, visualize it, believe it, then take action." You must act upon your dreams in order to turn them into physical reality.

Positivity:

Holding positive and loving thoughts will speed up the manifestation process. Manifestation is the art of positivity. Again, you must be aligned with positive and happy vibration in order to receive inspiration and ideas from the Universe. This goes along with the forgiveness step as well.

Doing what you love:

The truth is that we can do anything that we set our minds to. However, when you do what you love or what you are naturally passionate about, everything flows easily and smoothly for you. Someone who is naturally gifted with a great singing voice and dancing ability will be more successful as a singer and dancer than being a bodybuilder, for instance. Therefore, find something that you love, that you are passionate about, and you will naturally do extremely well in that field. If you don't like dogs or hate dogs, for instance, it will be very challenging and not pleasing for you to start a business in dog clothing. However, if you are a dog lover, and you are passionate about clothing and fashion, launching a dog clothing business can be very lucrative for you, and you can become very successful at it. Therefore, find something that you like and that you are passionate about and go for it.

Focusing on service:

Focusing on service is a big step in being successful. Ask yourself why you want to do these projects. The answer to that question has to be more than getting rich and making money. Your intention has to be to help others, to provide a service that will benefit many, to come up with a practical solution for society, and so on. Remember that you cannot trick the Universe and that your intention behind everything is what matters most. Therefore, you must focus on service, on helping others. When you focus on giving, you will automatically receive. This is a law of the Universe: the law of cause and effect. However, the ego will always push you to focus on yourself, and that is why it is important to purify your intention and truly be very honest with yourself about the reason why you can launch a business and do whatever you want to do.

The truth is that those who are really successful are those who focus on service: serving and making people happy and dancing with their artistic and musical talents, serving and bringing practical solution and easing people's lives with their talents in computers and software creation, serving others and making people happy joyous meals with their cooking abilities and their restaurants, serving others and uplifting and motivating others with a coaching business, and so on. The question you must always have in the back of your mind every time is this: "How may I serve?" When you desire to sincerely serve, the Universe will rush to your side and will provide you with countless helpful successful business ideas and inspiration.

For more information about how to manifest money and abundance, please check the back of this book for suggestions on manifestation books and meditations.

The Truth about Life
CHAPTER 22

THE SEVEN PILLARS FOR SUCCESS IN LIFE

"Truth allows you to live with integrity. Everything you do and say shows the world who you really are. Let it be the Truth."

–Oprah Winfrey

The most helpful thing that you can do for your life is to start working on yourself to become a better person, or I will say the best version of yourself that you can be. It is the best gift that you can give to yourself, and you will win on all levels. What do I mean by working on yourself? What I mean by this is doing inner work to become more loving, kinder, more forgiving, more peaceful, healthier, to quiet your mind and let go of the ego. However, you must decide and commit to being loving, to doing forgiveness work with all your relationships, including your relationship with yourself, and be healed from all painful or negative situations and traumas that you may have been involved in.

You must decide to purify your thoughts, to purify your lifestyle, and raise your vibration. No one else can do this for you, and only you can do the work. You may wish to check the healthy living chapter to see the things that you can implement in order to increase your vibration.

When you start working on yourself and 'do the work,' your vibration will start to rise, you will start to change; your environment will start to shift; all your relationships and your life situation will follow up and change as well; your life will change for the better. In fact, it is an unfailing law of the Universe. It is impossible for you to change from within without having your environment, your relationships, and your life situation change as well.

Let's take the example of someone who is working on himself to love others and who is committed to having love be the basis of his thoughts, words, and actions. Let's suppose that after two years this person ended up forgiving everyone who has ever hurt him, he heals his emotions and traumas, heals his body, becomes completely disease-free, and is practicing a very healthful lifestyle by exercising every day, eating healthy, meditating every day, turning off the television and spending his free time on creative projects, helping others and is truly living a life of integrity, joy, and happiness every day. It will be impossible for this person to become friends with those who practice hurtful behaviors such as gossiping, who use harsh or destructive language, who are constantly angry, who complain all the time, who drink alcohol, who eat unhealthy food and spend their free time watching drama programs on TV. It will be like trying to mix water with oil, which is immiscible. They cannot be friends because there will be a clash, and they will reject each other. This is because there will not be a vibrational match between them, and their vibrations will repulse each other. Therefore, in order to attract the good things in life such as love, wealth, joy, and loving relationships, you have to be a vibrational match for these things. In other words, if you want people to love you, you have to become a loving person. If you want to have successful, loving, and joyful friends, you have to embrace wealth consciousness, which starts within, and you have to become joyful and loving yourself.

In fact, there are people who live in a state of high vibration on this planet and who live in a state of the fifth dimension by being loving, keeping only loving thoughts, and live in the same way as Jesus lived thousands of years ago. They are rare, but there are few of them walking on this planet right now. However, in order to meet these people, become their friends, and be part of their small circle of friends, you have to work on yourself and become like them and live like them. There must be a vibrational match. It is a law. The millionaires live among themselves, and the poor live among themselves. The loving and honest people live among themselves, and the gossipers and angry people live among themselves and hurt each other. The healthy people live among themselves and support each other in their healthful lifestyle by exercising together and eating healthy foods, while those who do not practice a healthy lifestyle also live among themselves. This is because people who hang around together are of the same vibrational match: "Birds of a feather flock together"; "Like attracts like." Therefore, the most helpful thing that you can do for your life is to change yourself and seriously start to do inner work in order to raise your vibration.

Here are the steps that you can adopt in your life to change yourself from inside out, and I refer to these steps as the "Seven pillars of success in life."

The first pillar of success in life is self-love

I will say that self-love is like the basis or the foundation of success in life. From it are built the other pillars. Many of us do not love ourselves because we think we are not good enough or that we have made too many mistakes in life or because we are not as good as others. Society tells us that we have to be pretty, thin, rich, famous, own big houses, have successful and famous friends; otherwise, we are considered losers. Many think that they are not smart enough, not educated enough, not thin enough, not cool enough, not rich enough, not fashioned enough, not pretty enough, not a good enough parent, etc. Therefore, many do not accept themselves as they are and sometimes even despise themselves when they look at the mirror. Therefore, they do not love themselves.

Under the umbrella of self-love, there is self-worth. Sometimes we question our self-worth. "What are we worth?" we often ask ourselves. Do our lives matter at all? What are the points of all the things that we are doing: waking up, going to a workplace that we do not like, eating, sleeping, only to wake the following day and start over again and again? However, our worth does not come from our titles or career or physical beauty, but we are worthy simply because we are children of God. We are The Light, and this has nothing to do with our human behaviors, activities, ethnicities, accomplishments, or careers. However, every one of us has to rediscover this and live from that standpoint of unconditional self-love.

Under the umbrella of self-love, there is self-acceptance. Can we accept ourselves just as we are? Can we accept ourselves regardless of the mistakes that we have made in life? Can we accept ourselves regardless of our familial circumstances, our skin color, the events that happened in our lives? Can we accept ourselves and forgive ourselves for all our shortcomings in life? Self-love, self-worth, self-acceptance is the first pillar of a successful life, and it is something that can be cultivated and worked on.

The second pillar of success in life is self-confidence

Self-confidence is something that many of us struggle with. We lack confidence because we don't know who we truly are, and we don't know which talents and gifts we have and what we are truly capable of. Self-confidence and self-trust go hand in hand. I believe that lacking self-confidence comes from being lost. What I mean by being lost is not knowing your true essence, not recognizing your potentiality, not knowing what you are good at, forgetting who you truly are and believing the voice of the ego, believing what others think you are or say you are. To build your self-confidence, you have to somehow live according to your personal sense of integrity and your personal values, no matter what others say or think. Self-confidence comes from living your truth. I mean, by living your truth, doing what you know is true for you, what you know in your heart is best for you and not worrying about what people will think or what people may say and following your passion regardless of what society and others think.

Self-confidence is also built by knowing and doing what you are good at. If you are naturally good at something, it is best to cultivate that which you are good at, and you will excel in it. However, many people often set aside what they are naturally good at and instead try to force them to become something else or someone else instead of embracing their uniqueness and recognizing their natural talents. If you are naturally good in a field, when you polish that talent, things will naturally flow for you, you will excel in that, and you will enjoy it as well. Doing what you are good at will boost your confidence as well since you are good at it.

The third pillar of success in life is listening to your feelings

Listening to and honoring your feelings is one of the most important keys to success and happiness in life. The guidance system that we have been given to guide us through life is our feelings. The most reliable guiding tool that we have in life is our feelings. However, many do not know that. Oftentimes we try to suppress our feelings, push back our feelings, deny our feelings, or simply ignore our feelings. However, feelings are a guiding system and are telling you if something or someone is good for you or not. For instance, when you call a family member, and every time, this person says something that bothers you or makes you feel bad or uncomfortable, this simply means that the relationship is not good for you and that it is not in your best interest. It does not matter if you call that person a friend or a family member; your feelings are telling you the truth. However, we wrongly believe that we have to put up with some situations or people despite what our feelings are telling us. To know if a food, a person, a situation, a workplace, a city, or anything is good for you or not, the only questions that you have to ask yourself are the following (your answers to these questions are the truths, and these answers speak for themselves):

How do you feel after eating that food?

How do you feel in the presence of that person?

How does that person make you feel? Did the person do several things that angered you and the next time was nice and then did it again? If yes, why do you trust him or her again?

How do you feel when you are taking this course or class? Were you annoyed by the materials and wish you have something more enjoyable, more fulfilling to do? If yes, why do you keep forcing yourself to stay on that career path you don't like?

How do you feel in that relationship?

How do you feel in that workplace? Are you happy when you go there every day? Are you annoyed or even have anxiety at the thought of returning there? Your feelings are your answer and the most accurate guidance that you have been given in this world.

How do you feel in this city or country or place?

Success in life comes from listening to and honoring your feelings. It does require some practice and self-trust to live by relying only on your feelings instead of following people's opinions, the thoughts in the head, the societal norms, the expectations of others, or your family expectations. I personally started living by solely relying on my feelings a few years ago, and I can tell from my own experience that it has improved my life and enabled me to become more selective of my friends and the people that I allow in my life, and it has been life-changing and lifesaving.

The fourth pillar of success in life is mastery of self

Mastery of the self is an important step and key for a successful life. I will say that it is the cornerstone. Mastery of the self contains three components: self-realization, self-healing, and self-guided destiny.

<u>Self-realization</u> is when you come to know who you truly are and that you are not the ego and live in that blissful state of your true self. It is when you truly know your divine immortal nature, and this is only possible when you get out of the ego and live only in the present moment. This is done mainly through a daily meditation practice, which leads to calming the mind and being present in all your daily activities and taking the high road of love in all that you do. Self-realization is letting the ego die and letting love guide all your thoughts,

words, and actions. It involves self-awareness, embracing yourself just as you are, and truly living as your soul desires to live.

<u>Mastery of the self also involves self-healing</u>. The truth is that healing and living in the state of total health is one of the foundations of the pyramid of 'successful living,' and it is our divine birthright. How can you be at peace, live in joy, be happy, and enjoy your life if you are physically in pain or dealing with health issues? Can we call a life successful living if the physical body is experiencing discomfort and pain? Thus, self-healing is an essential component of successful living. Even dealing with health problems that some people call 'minor health issues' such as headache, heartburn, and seasonal allergies greatly interferes with peace and happiness because they are recurrent and annoying. There is no peace in taking medications just to breathe fresh air because of an allergy or having to take over-the-counter medications after eating because of heartburn. There is no peace in that. Keep in mind that we have come to believe that it is just a way of living and that it is normal, while the truth is that perfect health is our natural state; perfect health must become the new normal for successful living. We have to find a way to heal ourselves, and we cannot rely on taking medications as a way of living. Keep in mind that most medications have several side effects and that many medications actually cause other illnesses as well. Though medications can be helpful in the process of healing, we cannot rely solely on medications for our entire lives and must find a way to heal and care for our bodies, minds, and spirits. For more information on self-healing, please check the "Health and Disease" chapter of this book or look for the *Health and Disease and Prevention* book that I will be publishing in the near future.

THE TRUTH ABOUT LIFE

The fifth pillar of success in life is self-guided destiny

This journey that we call life is your journey; it is a gift that you have been given to rediscover yourself, to embrace your uniqueness, to create the life that you desire, and to enjoy. It is your journey, not anyone else's journey but yours. It is not the journey of your children, of your parents, of your teachers, of your family, of your lover, of your spouse, of your celebrity idol, of your spiritual teacher, of your counselor, of any model of society, of any guru, of any religion but 'your journey.'

You must live your self-guided journey, or I will say your 'God-given destiny,' whatever that may be for you. Only you can live your life and fulfill your destiny. You have to discover and fearlessly embrace your destiny, your purpose, your call. It may not make sense to anyone; it may seem absurd to others; it may not be in alignment with your religious background, but when it is your destiny, your call, your most profound desire, you will know it, and you will feel that in your heart. You must follow your own path.

You must trust yourself enough and follow your path. You must be strong enough to do what matters the most for you and to do what brings you joy. You must be courageous enough to go outside the box. You must become brave enough not to let others' opinions and lack of understanding deviate you from your destiny. You must become bold enough to stand tall and follow your life's path because you have one, one that you planned before your birth. You must become audacious enough to jump and take steps in the direction of your life purpose. You must have confidence enough in yourself to face the nonunderstanding of others and to face the challenges that may be ahead on the road. Yes, it is your life; it is your time. It is your self-destiny. The guide, the guideline, the how, the messiah, the way, the torch, the divine law, and the instructions are already outlined within you, in your heart. It is your journey: go for it.

The sixth pillar of success in life is loving relationships

Can we call a life a successful life if you do not experience love in your relationships? Can we call a life a successful life if we go to work and are in the company of people who hate us and with whom you don't get along? Can we call a life a successful life if you come back home, and you do not find peace because of tension with your family, partner, spouse, girlfriend, or boyfriend? Can we call a life a successful life if we do not have peace with our relatives and if there are constant tensions and arguments? What peace is there in that? It is a no-brainer that 'loving relationships' is one of the pillars of success in life. However, the truth is that the love or the strain that we experience in our relationships is often a reflection of how we feel on the inside. Loving every part of yourself and being at peace with yourself somehow will help to improve your relationship with others. Forgiving others and accepting them as they are will ease your relationships as well.

The biggest element that creates tensions in relationships is 'judgment.' Judgment is when we criticize others or judge them or their actions as being wrong or inappropriate, which automatically leads to anger. Judging others creates pain and strain in relationships. However, the truth is that "Judgment always follows judgment." What does that mean? It means that when you judge someone, the person will not be happy with you, will be angry, and the person will judge you in turn. It is often automatic. *"Do not judge and you will not be judged. Do not condemn, and you will not be condemned. Forgive, and you will be forgiven"* (Luke 6:37). Therefore, if you don't want to be judged by others, then do not judge anyone. The truth is that someone's path is intrinsically embedded in their soul. All paths lead to God, which means that we will all return and merge to the Light sooner or later, no matter the path that we adopt during our earthly incarnations.

Therefore, judging someone is rejecting them, and it is similar to judging the path that their souls have chosen as wrong, and it's why when you judge someone, the person is hurt and will end up judging you as well.

Judgment is the biggest destructive weapon in relationships, and one of the best things that you can do for yourself is not to judge anyone. Do not judge your parents. Do not judge your coworkers. Do not judge your family-in-law. Do not judge your neighbors. Do not judge your friends. Do not judge your enemies. Do not judge your brothers and sisters. Do not judge people of other ethnicities. Do not judge strangers. Do not judge yourself. Do not judge anyone or anything. Do not say, "Why do they act like that? Why are they wearing this? Why are they homeless while they have both hands and can work? Why are they obese? Why are they stealing? Why are they lying? Why do they look like that?" etc., because you don't know their souls' paths. You don't know their familial background. You don't know what happened in their lives. You don't their past-life incarnations. You don't know which lessons they are working on. You don't know what is happening in their minds. You don't know what they are going through in their lives. In fact, you don't know anything about them. Therefore, do not judge them.

Only the ego likes to make quick judgments. The truth is that human beings are very complex beings with a blend of emotionality, mind, spirit, memories from childhood, memories from past lifetimes, and with a complexity that we cannot fully comprehend with our minds. However, judgment happens when you see someone, and you say, "I know him; this is John. He is black, or he is white. He is fat. He always does this or that. He is strange. He is this or that." Judge not because, in truth, you know little about them, and they are more than you can see. They are more than their acts. They are more than anything that you think you know about them.

Yes, loving relationships are a pillar of a successful life, and loving relationships come from loving yourself, forgiving yourself and all others, practicing compassion, and not judging anyone for anything.

The seventh pillar of success in life is trusting Source/God as your primary point of wealth, peace, and happiness

Your religion, belief, spirituality, or atheism does not matter because everyone, no matter their religious background, has guardian angels assigned to them. The truth is that these guardian angels are unconditional loving beings and do not have ego; therefore, they will never condemn you, judge you, be angry with you, or have their feelings hurt by you. Also, they also operate from the place of free will; therefore, they will only give suggestions and guidance, and then it is up to you to decide if you want to follow this guidance or not. They are here to guide and support. I have discovered from my personal experience that the most painful and difficult periods of my life were when I believed that I was alone and abandoned in life. Developing a relationship with the Divine is perhaps the most helpful thing that you can do in your life. It does not matter if the Divine for you represents God, Jesus, Allah, the Buddha, Mother Mary, or your angels. The truth is knowing that Universe/God is here to support you, and trusting in the support is very powerful, comforting, and life-changing. However, this comes with the understanding of what the Divine is, the very loving nature of the Divine and its unfailing and unconditional love and support.

We suffer, and we feel isolated and alone when we get the truth messed up and when we misunderstand who or what God is and start to

THE TRUTH ABOUT LIFE

believe that Universe/God may be angry about us, punish us, or abandon us, or we believe that we are alone. Unfailing trust in the Divine is one important pillar of success in life. Trust that Universe/Source/God is here for you, unconditionally loves you, is here to support you, and only wishes for your peace and well-being. What often happens is that we ask questions like these: If God is good, loves me, and supports me, why do I have this disease? If God is good, why I am suffering? If Source is benevolent, why am I poor, even though I am a hard worker and have worked hard all my life? If Source is good, why was I born in this family and am disadvantaged in life? If God is good, why I am suffering, even though I am a good person? If Universe is benevolent, why has this or that happened to me or my prayers are not being answered? etc. However, God is not the cause of your health issues as explained in the health and disease chapter.

The truth is that you have chosen some of your life circumstances and your parents before your birth in accordance with what you want to learn, and God and your angels are the ones guiding you to ease your path and to make the experiences that you have chosen to be better, easier, smoother, and more joyful for you. The truth is that abundance, like many things in life, follows some laws such as the law of attraction, and it is something that everyone can learn in order to switch their situations around and manifest wealth. Therefore, understanding the principles of creation and the truth about life will help you to trust in the Divine. Trust, belief, and faith are the platforms from which miracles spring forth along with essential elements for a successful life.

Trusting in the Universe as your primary Source for wealth, health, and love is an important pillar of success in life. However, before you can trust someone, you must know that person; furthermore, you must develop a relationship with that person. For instance, it was by developing a relationship with my angels that I discovered to my astonishment that again and again, they are consistent in what they say, always loving and unconditionally supportive. Getting in touch with the Divine and developing a relationship with God, your guides, and guardian angels is an essential pillar of success in life.

ACKNOWLEDGMENTS

I'd like to thank my sister Tutu for her love and support, for standing by my side, and for believing in me. Many questions included in this book come from our passionate conversations and from times spent discussing various topics. I am grateful for your presence in my life, dear sister.

To my book editors Tessy and Ken; for their kindness, patience, and amazing work. By making structural adjustments and grammatical corrections, they have made this book more agreeable for all the readers.

To my friend Sara Troy; for her kindness, unconditional support, and for introducing me to many people. Thank you, Sara, for helping to record these prayers and affirmations that helped me to be where I am today.

I would also like to thank my daughter, Brielle, for being part of my life and for all the joy and blessings that she has brought into my life.

I would like to thank my heavenly team: Jesus, Saint Augustine, Mother Mary, the Buddha, Saint Germain, Yogananda, Athella, and the 12 Archangels (Archangels Michael, Uriel, Gabriel, Raphael, Chamuel, Raguel, Jophiel, Metatron, …etc.) for their help beyond words and for their unconditional love. This book would not have been possible without their help and support. Thank you, my angels! Thank you, heavenly guides!

POSTFACE

I really hope this book brings clarity and peace into your life. Writing this book has really been a great blessing and joy for me. May you be blessed now and forevermore.

REFERENCES

For wealth and abundance, please refer to the following books:

Think and Grow Rich,
By Napoleon Hill

The Abundance Book,
By John Randolph Price

The Power of the Subconscious Mind,
By Joseph Murphy

For wealth and abundance meditations, please refer to the following meditations:

The 6 Guided Phases Meditation,
By Vishen Lakhiani

For more information about the ego and the egoic mind, please refer to the following books:

The Power of Now: *A Guide to Spiritual Enlightenment,*
By Eckhart Tolle

ABOUT THE AUTHOR

Dr. Lilly Koutcho

Dr. Koutcho is a pharmacist, a writer, and a spiritual counselor. She works with Archangels and ascended masters for the purpose of healing, teaching, and service. She holds a master's in neuroscience, a master's in Business Administration (MBA), and a doctorate in pharmacy. She is the author of several books and the founder of the nonprofit organization Light & Love for Africa.

To learn more about Dr. Koutcho and her work please go to: www.integrityhealing.org

Made in the USA
San Bernardino, CA
07 March 2020